OF SCIENCE AND GOD

Science is overwhelming. Why believe in God?

MICHAEL LEWIS WHITE

Of Science and God
Science is overwhelming. Why believe in God?
All Rights Reserved.
Copyright © 2022 Michael Lewis White
v4.0

The opinions expressed in this manuscript are solely the opinions of the author and do not represent the opinions or thoughts of the publisher. The author has represented and warranted full ownership and/or legal right to publish all the materials in this book.

This book may not be reproduced, transmitted, or stored in whole or in part by any means, including graphic, electronic, or mechanical without the express written consent of the publisher except in the case of brief quotations embodied in critical articles and reviews.

Gnostic Press

ISBN: 978-0-578-25476-0

Cover Photo © 2022 www.gettyimages.com. All rights reserved - used with permission.

Outskirts Press and the "OP" logo are trademarks belonging to Outskirts Press, Inc.

PRINTED IN THE UNITED STATES OF AMERICA

*Dedicated to my father
Lewis McKinley White Jr.,
August 6th, 1932 to August 6th, 1981*

He showed his love by being strong for others. Born dirt poor and raised in a situation where while there was always food on the table, the food was never enough to completely satisfy his hunger. He went on to found his own Interstate Highway signing company. When his oldest daughter Michelle got married, he took a trip out to Colorado and got to know her new husband Richard. Later he told Michelle "So long as there is breath in that man's body, you will never starve." It was the greatest compliment he ever paid anybody. The remark says a lot about the kind of man he was and while he lived no one he ever loved went without anything they truly needed.

Table of Contents

Preface..i

Science Section..1
    Foreword to Science Section 3
    Relativity..7
    Relative versus Absolute Space......................................19
    The variable speed of light and Time............................39
    Cosmology 58

Humanity and the Divine 73
    Belief..75
    Evolution 77
    Human spirituality..85
    God...94

Religion..105
    The Creation of Religion..107
    Christ and Christianity...109
    Islam 125
    Judaism 134
    The Eastern Religions 140
    Prayer...142
    The Abuse of Faith...149
    Rational Faith in a Technological World 156
    Judgment 163
    The Second Coming 171
    The Challenge to all faiths..182
    The Ultimate Fate of Humanity?................................184

Preface

Science is science and the belief in God is ultimately a question of faith. Why am I mixing the two in this book? Many people think science disproves God. That is not the case. People don't understand that science is a human endeavor with all the flaws of human nature, built into it. Science and technology manifest themselves in the tools with which they provide us, not in the values they instill. Science and technology don't solve or address our common need for identity or help us find meaning in life. Science and technology are cold and indifferent to the human condition.

We live in a world of exploding technology that is rapidly breaking down the old cultures and building in its place a new world-wide cultural landscape that worships technology and money. This culture is best summed up by the bumper sticker, "He who dies with the most toys wins." This new culture breeds nihilism and there is a growing belief that evidence of God and the supernatural does not exist within the realm of human experience, that such things are falsehoods and are simple superstitious beliefs.

Pope John Paul used to call it the culture of death and it has a cost in society. Mental health professionals estimate that at any given time fully twenty percent of the population is suffering from some form of mental illness, the most common being anxiety and depression. While anthropologists find very little or no mental illness in primitive societies that still live as hunter gatherers as humanity originally evolved. We live in the age of Prozac and video games. In many respects, it is poor trade from our contended past to the common angst of our technologically driven society with all of its distractions. The historians Will and Amy Durant made two predictions in their book "The Lessons of History". One that communism would collapse and two that society would run back to Old Mother Church. Let us hope we do it with eyes wide open and not in fear or superstition.

I believe in evolution and an infinite and eternal universe, a universe with no beginning and no end, in size or time. A universe not necessarily created by God. Yet I have a firm and unshakeable belief in God. Why? Because there are things in life that science cannot and never will be able to explain. I seek to establish a rational and scientific argument for God. I am writing this book to drive a stake in the heart of the argument that Charles Darwin's evolution proves there is no God. Understand something and understand it well, it is my understanding of science and how inadequate it is to explain certain things that drives my belief in human spirituality and God.

The primary purpose of this book is to get people to think. If you are thin skinned or your belief in God is superficial this book may make you angry at me for daring to suggest alternatives to what you see as the truth. My purpose is not to annoy, but rather strengthen belief in God and the doctrine of Christ. I try not to use inflammatory rhetoric, but rather strengthen what I see as the positive convictions of a great many people through informed debate. Please read this book with an open mind.

Science Section

Foreword to Science Section

If you do not like relativity or general science you can skip right to the section on humanity and the divine. This is followed by a religion section with a discussion of the various religions, the second coming and the ultimate cosmic fate of humanity. I warn the reader of something. The science get very advanced at points in the discussion of the issues involved. If you get lost don't give up on the rest of the book. Anyone with a high school education can understand the sections on religion. While in some respects the science section is an advanced discussion. I am a great believer in the old saying "If you can say it in plain English, you don't know what you are talking about". The first and last chapters in the science section on relativity and the universe fall into this realm. They should be understandable to the general reader. Feel free to skip around parts of the science section if you want.

Albert Einstein's theories of relativity are generally considered to be the greatest scientific and intellectual achievement of the twentieth century. Our entire picture of the universe is shaped by Einstein's theories and the Big Bang theory. I think the big bang theory is crap and further think relativity needs modification. Who am I? Michael Lewis

Michael Lewis White

White, a nobody. What's my advantage I am a super genius. I maxed the online Mensa IQ test. I read Einstein at an early age and have always disagreed with some of Einstein's basic underlying assumptions since that time.

My assertions in this book are my own. I have astronomically high logic and pattern skills. I have a history of maxing logic and analysis tests. My scientific ideas are logical deductions from the simple premise that there is no such thing as a universal constant. All those inert atoms don't have common agreed to truths, that apply in all circumstances to all the variations and configurations. Our galaxy is collapsing towards a massive stellar formation composed of millions of galaxies. All the constants of physics are in flux. My purpose in writing this book is ambitious. I seek to prove an expanded subset of relativity, while arguing against curvature of space. Like I said it is an ambitious agenda.

Albert Einstein lived in a time dominated by the achievement of Newton who described the natural order of the planets and physics in a succinct fashion, just as Darwin explained evolution as the natural order of the biological world. This was at a time when universal constants such as mass and G were thought to be truly universal. This was a direct result of Newton's work. In special relativity in 1905 Einstein proved mass was relative to speed and later in general relativity proved time was relative to mass and speed. In general relativity Einstein was attempting to prove that the speed of light was a universal constant and to date his argument is believed to be a success because of the accuracy of Einstein's predictions regarding atomic time. These are extremely accurate and have stood up to test after test. I will explain why this math is wrong, self-deceptive, and will provide an alternative math to test my contention. I contend Andrei Sakharov's math for atomic time, does not explain the physical conditions inside the atom. Some people think math is the only truth. But it can be wrong or can be used to fool oneself. Math can be sophistry. When math doesn't match reality, it is pure sophistry.

I started out not believing in Albert Einstein's theories of relativity when I first read him at age 10. I read the final edition of his book Relativity. In this book Albert Einstein, went to great lengths to explain how an astronaut in a rocket ship traveling at ninety percent of the speed of light would if given the proper tools, would always measure the speed of light of a laser beam overtaking the rocket at exactly 299,792 kilometers per second, due to the slowdown in time and contraction in lengths. That statement is true enough, but the weakness in his argument was always this: how would the astronaut measure a beam of light traveling in the opposite direction? I think relativity, with all its vast assumptions needs to be modified and seek to shed some new light on the debate.

Albert Einstein's theories of Relativity have evolved since he first described them over a hundred years ago and are no longer logically coherent with Einstein's original arguments as put forth in his own words in the first edition of his book "Relativity". People are working with the latest reading or observation and do not bother to resolve these observations with Einstein's original logic. My advantage is that I have read Einstein in detail several times and I am dealing with his logical construction and not trying to fit the observations into the theory without criticism.

My challenge to relativity is to the curvature of space. I accept relativity's basic underlying premise that things expand and contract according to changes in the speed of light. However, I assert that you cannot hold on to the speed of light as a constant unchanging measure of space. Space, be it the space between planets or the space between the basic particles of the atom, must be calculated in terms of absolute space. I do not believe space is curved. I do not believe space is metrically elastic. I will explain this in some detail, but will argue different aspects of my thinking in different parts of this book. So please read the entire book in the order presented. Please read the rest of the book if you get lost in the science section.

I warn the novice that the science and logic in this section of the book is quite complex and difficult to follow at times. If you as a reader get lost, slog on through until you reach a part of the book you understand. Finish the entire book before going back to reread selected parts. It is going to take a decade or more for scientists and astronomers to prove or disprove everything in this book. If you get totally lost in the discussions in the science section don't give up on the rest of the book. The religious sections of the book are very fairly straight forward and simple. Though they may be controversial.

Relativity

"It would take only one, if he were right"
a remark of Einstein when told the Nazi's were publishing a book "A hundred against Einstein".

What is Relativity?

There are two theories of relativity. The first is Special Relativity and the second is General Relativity. Taken together they are collectively referred as relativity. Relativity is hailed as the greatest intellectual achievement of the twentieth century. Albert Einstein developed his theories of relativity from the Michelson and Morley experiment and the Lorentz transformation. A quick explanation of the experiment is in order.

At the time of the experiment light was not understood as a particle. The wave particle duality did not yet exist in scientific thought and light was thought to be waves propagated through a fluid known as 'Aether'. A fluid that was present in all space and effortlessly passed through all matter, whenever that matter moved. The Michelson and Morley experiment was an attempt to measure our motion through that 'Aether'.

The entire experiment is on swivel. It splits and recombines a beam of light into two beams of light at a ninety-degree angle to each other with the help of a slivered mirror. The theory being that the motion of the earth would cause the split beams of light to be out of sync when they recombined and by measuring these bands of interference one could measure the earth's motion through the 'Aether'. It didn't work, no matter how you turned the experiment, there were no bands of interference, no discernable motion at all.

Scientists were faced with apparent proof the earth was motionless at the center of the universe, as was thought in ancient times. To resolve these difficulties, a mathematician named Lorentz produced a mathematical explanation of why the experiment showed no discernable motion, based on the premise that the experiment contracted in the direction of motion. Albert Einstein took this formula and came out with the special theory of relativity in 1905, which not only held that there was a contraction in lengths, but as a body of mass contracted in the direction of motion. It also increased in mass by the same formula. There is not one shred of proof of the contraction of lengths due to motion anywhere in all of science outside of the Michelson and Morley experiment, but the increase in mass however is verified every time they use a particle accelerator. Niels Bohr had built a small particle accelerator a hundred years ago. It worked exactly as Einstein had predicted.

The second part of relativity General Relativity, which was developed a few years later, maintained that the speed of light could never be measured as anything other than 299,792 kilometers per second in a vacuum anywhere in the universe or on any celestial body in motion. This was due to his contention that time itself slowed down or increased as space contracted or expanded as the speed of light changed. In general relativity Einstein maintained that motion and gravitational potential were one and the same thing. Einstein also maintained the universe was one coherent entity and that the fabric of the universe was such that nothing could exceed the speed of light within its domains. Relativity requires that space be relative not absolute as Newton maintained. General relativity was intended to be replacement to the theory of gravity. In general relativity there is no gravity only the curvature of space and motion through that curved space.

Relativity is considered to be correct because of the repeated verification of motion and gravity's effect on atomic time, but relativity also requires that space expands or contracts due to the influence of other bodies of mass. This aspect of relativity is called the curvature of space or the Metric Elasticity of space.

The contraction in lengths in the direction of motion has no demonstrable effects in a particle accelerator. At 99 percent of the speed of light an electron would not be spherical but rather would be flat as a ribbon and shaped somewhat like two plates or very shallow bowls placed on top of each other. It would be very narrow at the edges on a right angle to the direction of motion and broad in the middle. This would alter the nature of particle collision with the spherical particles at rest in the bubble chamber and should produce some noticeable effect. The impact of a contracted particle with the atomic particles in the bubble chamber would take on the characteristics of a flat surface hitting a sphere. It does not and there is no evidence of a contraction in lengths due to motion anywhere in science, outside of the Michelson and Morley experiment.

So, what is the explanation of the Michelson and Morley experiment? Why did the experiment show no motion? As we have noted before the experiment is predicated on the assumption that light was a wave propagated through a fluid called 'Aether' occupying all space. This idea has been discredited by proving light is a particle as well as a wave and needs no fluid for its propagation through space. Albert Einstein did that in a paper on the photo-electric effect. The explanation of the experiment is simple: light rotates as the gravitational field in which it is traveling rotates and there is no contraction in lengths.

Something that supports this contention is that the measurements of the moon's distance from the earth by laser beam. The Apollo astronauts put a laser reflector on the surface of the moon and scientists get very accurate measurements of the distance between the Earth and the moon by measuring the time it takes for light travel to the moon and back. The first time they tried this they missed the laser reflector by a distance of ten miles. This is the exact distance the earth gravitational field rotates in the one second it takes for the laser beam to transit the earth's gravitational field and enter the influence of the moon's gravitational field which doesn't rotate.

What most people do not understand about general relativity is that the speed of light is held to be variable, but variable with a constant relationship to metrically elastic space and relative time. At 40,000 feet light is moving at one tenth of one percent faster than at sea level. This change in speed is accompanied by changes in time and the metric elasticity of space. At 40,000 feet clocks run faster and any set of measuring rods would expand. So that if you were to measure the speed of light at forty thousand feet, the speed of light would still be measured at 299,792 kilometers per second. Because time speeds up and space expands. The speed of light changes but so too do time and space. Remember that for general relativity to work, the speed of light must be variable, but so too are time and space in a constant relationship to the variable speed of light.

For general relativity to work all three of these things, the speed of light, time and the metric elasticity of space must always stay in balance. Space must expand or contract and time must speed up or slow down, such that locally the speed of light is always and can only be measured at 299,792 kilometers per second. Einstein said that was exactly what would happen time would speed up or slow down and that space would contract or expand and because of this the speed of light could never be measured as anything other than 299,792 kilometers per second in a vacuum. Time has been shown to be variable exactly according to Einstein's predictions. But the contraction and expansion of space is still considered conjecture and if space is not metrically elasticity, does not expand and contract, the mechanics of relativity fall apart.

The speed of light slows down near the surface of the sun. If you use a radio telescope as a radar device and bounce radio waves off a planet as it begins to pass behind the sun, the planet appears to jump thousands of miles out of its orbit. This is directly attributable to the slowdown in the speed of light near the sun. Einstein's predictions about how motion and gravitation potential effect time have been proven correct countless times. These are two of the three things that must always stay in balance for general relativity to work. Leaving only the metric elasticity of space. Space has to contract the exact same percentage that the speed of light slows down near the surface of the sun or general relativity is wrong.

Einstein was convinced everything was geometry. Why? Because according to relativity and his thinking if you put a disk or sphere in motion it would contract in direction of motion distorting Pi. This is how Einstein came up with his curvature of space, motion distorting pi. His argument was: if motion causes measuring rods to contract in the direction of motion, a uniform set of measuring rods measuring the circumference of a spinning disc would be shortened in the direction of motion changing the relationship in between the number of rods measuring the radius and those measuring the circumference of disc altering pi. This is the premise scientists say that they confirm with atomic time.

The curvature of space has been transformed into the metric elasticity of space and is a difficult concept for the average person to understand. Don't feel bad the average scientist does not understand it either. I will try to simplify it for you. According to current theory the closer we move to a large mass the very atoms shrink, and the field forces that surround the atom shrink, creating more room for matter. One must understand that Einstein originally intended the curvature of space to be a replacement for gravity and the metric elasticity of space is a fancy argument that all things expand into contracted space.

This theory of metrically elastic space conflicts with Einstein's original explanation of the curvature of space. The explanation of the curvature of space with which most people are familiar is that of a rubber sheet or some other surface depressed by a ball bearing. This is the way most text books explain the curvature of space. I am not sure if Einstein came up with this explanation of the curvature of space, but it is a very good example and conforms to Einstein's original explanation and ideas about the curvature of space. Draw a grid of parallel lines on such a surface when it is flat and level and you will see when it is depressed by the ball bearing the distance between the lines of the grid expand the closer you get to the ball bearing. It does not contract as theorized in the metric elasticity of space.

Stay with me here, as we get into the curvature of space a little more deeply. Like I said Einstein's curvature of space was based on a rotating disk. Einstein said the rotation of the disk would cause a uniform set of measuring rods measuring the circumference, or edge, of the disk to contract in the direction in motion. This contraction would alter the relationship between the rods measuring the diameter of the disk and the rods measuring circumference altering pi. The further out on the disk you move and the faster the disk spins the more the rods measuring the circumference would contract altering pi. This feature of the curvature of space is synonymous with the gravitational potential of an area of space. On Einstein's rotating disk the longer rods are in towards the center, while the shorter rods are on the edges. Metric elasticity of space works in reverse of Einstein's rotating disk. The longer rods are out on the edges and the shorter rods are in toward the center. This perception of space is what physicists say they confirm with atomic time.

The contraction of space in the metric elasticity of space is why black holes are said to be sucking space into them. If you take Relativity's proposition that nothing can exceed the speed of light in contracted space to absolutes, combined with its slowdown in time and the contraction in lengths, it means nothing can ever be sucked into a black hole. Because the closer you get to the event horizon of a black hole the more time slows down and space contracts. This means that any physical body of mass being sucked into a black hole must continually slow down until it stops at the event horizon or it will exceed the speed of light in contracted space. Because gravity slows down light and space contracts with a corresponding slow down in time it can be mathematically demonstrated that a light year is less than 100 miles in absolute space near the event horizon of a black hole.

If you understand Einstein's original explanation of the curvature of space, it raises an interesting question. If space is metrically elastic, exactly how does metrically elastic space bend star-light? The standard answer is that light follows a straight line in curved space. Straighter than straight as they used to say back in the thirties and forties of the last century. A straight line is the shortest distance between two points and the claim about light following a straight line in metric elastic space doesn't stand up to detailed analysis.

Einstein maintained that the measuring rods were longer closer to the center of the rotating disk. Because the slower you rotate the less the rods measuring the circumference would contract for that area of the disk and you need fewer measuring rods the closer you got to the center of rotating disk curved space. Einstein said this expansion of space was how large bodies of mass bent star light. That light was following a straight line in curved space. It was following the path of fewer measuring rods, the shortest distance between two points in curved space.

The metric elasticity of space does not work on those assumptions. Space contracts not expands as you move inward to the center of an area curved space thus increasing the number of measuring rods you need to measure the path, so light should bend outward along the shorter path. That is the shortest distance between two points. A straight line is still the shortest distance between two points and space is not metrically elastic.

Let us examine this proposition of contraction and expansion using measuring rods one meter in length and a distance of a thousand meters in absolute space. Remember a straight line is the shortest distance between two points. The path defined by the fewest measuring rods.

In a normal world, it takes 1,000 one-meter measuring rods to measure a distance of 1,000 meters. In the metric elasticity of space, space contracts as you move in toward a large stellar mass. Light slows down near a body of mass like the sun or a blackhole. Let us say that the speed of light slows down 10 percent, the corresponding change in space gives you a new measuring rod of 90 centimeters and you now need slightly over 1,111 measuring rods to measure 1000 meters in absolute space.

Einstein contended space expanded near large masses. Let us say for the sake of argument that space expands ten percent near a theoretical large mass. A corresponding change in space gives you measuring rod of 1 meter and ten centimeters and it now takes slightly less than 910 measuring rods to measure 1000 meters in absolute space. A straight line is still the shortest distance between two points. It is 910 to 1,111 measuring rods Clearly the straighter path, the path with the fewest measuring rods is composed of the path with the longer rods. According to the metric elasticity of space and Einstein's original explanation of the curvature of space works large bodies of mass should bend star light outward. When the sun bends star light inward it is bending it along a longer path. Space is not metrically elastic.

Albert Einstein intended the curvature of space to be a replacement for gravity. He sought to explain the attraction of mass as motion based on the curvature of space. The metric elasticity of space turns Einstein's original arguments for curvature of space inside out and upside down. Einstein who once remarked "Now that mathematicians have gotten a hold of relativity, I hardly recognize it anymore." failed to catch the error for thirty years. When Albert Einstein published his theory of General Relativity and for decades afterwards, he continued to insist the curvature of space and gravity were mutually exclusive-that there was no gravity, only the curvature of space. You could not have both. I understand Einstein's original arguments and assert for you now, there is no such thing as the curvature of space-only gravity because I read and understand the logic of Einstein's own explanation of that contradiction.

If you want to understand this argument, get a copy of Professor Gamow's book "Gravity". Professor Gamow was a contemporary of Einstein and a genius in his own right, with that rare gift to explain complex subjects in plain down to earth language. In this book Professor Gamow not only explains both Isaac Newton's theory of gravity and Einstein's curvature of space, but also the logic behind Einstein argument that gravity and the curvature of space were mutually exclusive. These arguments are basically the same as the arguments set forth in this chapter. George Gamow personally knew Einstein and discussed relativity with him. Now that Einstein is dead Professor Gamow's book and the first edition of Einstein's book "Relativity", is as close as we will come to too getting it from the horse's mouth.

One must understand there is a difference between the metric elasticity of space and space. Space is a void, which we arbitrarily divide up into units to quantify. The size of objects is dependent on the size of the atoms of which they are composed. Saying atoms change in size according to the speed of light is not the same as saying space changes. To modify relativity, you must be able to measure the changes in the size of these very atoms in relation to absolute space.

This is quite a challenge and is one reason why Einstein's relativity has endured virtually unchanged for over a hundred years. Remember any modification of relativity requires that you prove that things expand and contract according to changes in the speed of light, while at the same time proving space as absolutely independent of the size of things. That is what I am endeavoring to do in this book. I will try to do it clearly and concisely so you can follow me and understand the arguments and assertions in this book. Wish me luck and read on.

Einstein himself late in life abandoned his work on the unified field theory which was the ultimate culmination of all his work on relativity and he then worked on and theorized about gravity. Because Einstein worked on gravity and without renouncing the curvature of space, gravity and the curvature are thought to coexist. They do not. There is no such thing as the curvature of space, only gravity.

Einstein was human and did not openly trash his own theories of relativity, which stand up to every electromagnetic test and experiment you can think of. All electromagnetic phenomenon is relative, but gravity works in absolute space with relative time. In a successive chapter I will attempt to prove time is dependent on gravity as it effects on the energy of the particles and not the curvature of space. I am taking Einstein at his word.

Confronted with the collapse of his logic construction, and its apparent proof at the same time Einstein retreated. He rewrote his book on relativity and attributed the changes in time to gravity, without abandoning the contraction in lengths. He went through several editions book "Relativity". Then finally he came up with his theory of gravitational waves offering a way to prove the existence of gravity. He rewrote the final edition of his book "Relativity" including his theory of gravity waves while deleting his detailed explanation of the curvature of space all together. I believe Einstein deleted the detailed discussion of the curvature of space because he knew his ideas about the curvature of space were wrong. I challenge current scientists to read the first and last editions of his book and see what they think.

Again, Einstein deleted his explanations of the curvature of space and developed his theory of gravity waves offering a method to prove the existence of gravity. I believe he did this based on his original contention that gravity and the curvature of space were mutually exclusive. He deleted his description of curvature while at the same time offering a method to prove gravity. Gravity waves have since been observed and proven to exist, but because Einstein worked on and theorized about gravity without openly dismissing curvature. The two are thought to coexist and be interrelated. People have forgotten Einstein's exhortations that gravity and the curvature of space were mutually exclusive. Scientists who say gravity and the metric elasticity of space coexist, have done so without Einstein's approval. This is a mistake that grows out of the hero worship of Einstein.

This is because Einstein never openly denounced curvature. Scientist are not ready to abandon it. Einstein was human and one can hardly blame him for not calling his most famous work wrong. Scientists in worshipping the ground Einstein walked on, are ignoring the practical down-to-earth implications of what he actually said. This is a mistake. The error seriously threatens the advancement of scientific thought and theory.

Relative versus Absolute Space

"Space is absolute"

Sir Isaac Newton

Isaac Newton defined space, mass and time as absolute things, unalterable as physical manifestations in reality which were unchanging features of the universe and he mathematically described the motion of heavenly bodies as a function of gravity. Albert Einstein came along and proved mass and time were relative. Then Einstein in 1917 came out with General Relativity by which he attempted to prove there was no such thing as gravity-only the mutable nature of space, as mathematically described by the mutability of Pi. In this effort he dispensed with Newton's absolute space, Einstein argued that space was not independent of its measurement, that the very tools you use to measure space were relative to the speed of light and motion, therefore space must be mutable.

Today's scientist is faced with the very real prospect that all the tools by which they measure space are relative while space itself is absolute. Both physicists and mathematicians are faced with the prospect that numbers are abstract concepts that represent real unalterable measurements of an absolute physical reality and they must devise methods of using these relative tools and instruments to discern these values in absolute space. At current this need is not being addressed due to the mathematically correct predictions of relative effects of space and motion upon time according to the mutability of Pi. The formula for atomic time assumes there only form of space, relative space! No attempt is being made at present to work with absolute space in the calculation of atomic time. This may be a serious mistake. Does Planck's constant work in absolute space? Before we tackle that question let us examine the contrasts between absolute and relative space as defined by Einstein and Newton and as they effect measurements in physics and astronomy.

Newton and Einstein have two fundamentally different concepts of space. Newton defined space as absolute and did not specify any specific tools for measuring space. Newton's space is a void, an abstract value we arbitrarily define and divide up in to set and fixed increments as a means of quantifying and understanding motion of bodies of mass though it. Einstein thought space was relative to the changing size of all material bodies or objects. As size of objects expanded or contracted, space also expanded and contracted. This was because all things were composed of atoms which expanded or contracted uniformly with the changes in the speed of light. To Einstein space could not be independent of the tools used to measure it. Those tools were relative to the motion and the speed of light.

There is a method in Astronomy to measure the physical size of atoms in areas of space with a different speed of light. You measure the displacement of the spectral bands of the hydrogen atom. They contract and/or expand depending upon the height or frequency of the electron orbits. The displacement of the spectral bands of the atom is based on Planck's constant. If one can prove the atom changes in size depending upon where you measure it, then you can prove space is absolute and that the size of the atoms is relative.

I submit you can do this by measuring the relative frequency and height of the electron orbits or the displacement of the spectral bands of the hydrogen atom on the surface of the sun where the speed of light is different and comparing them to the spectral bands of hydrogen atom here on earth. According to the following formula:

$$(V_1/V_0) = (C_1/C_0)$$

Wherein: V_1 is the displacement of the spectral bands on the surface of the sun, V_0 is the displacement of the spectral bands here on earth, C_1 is the speed of light on the surface of the sun, and C_0 is the speed of light here on earth.

One should note Planck's constant is an increment of a given electron shell. The electrons orbit the atom as one coherent wave spanning several electron shells and Planck's constant is a basic increment of that wave. Planck's constant is a measurement in absolute space. If space were relative, you could not measure a difference between atoms here on earth and on the surface of the sun. Both would be the same constant size. There is no test, no measurement you could take to measure the difference in the size of atoms of the same isotope of a given element here on the surface of the earth. This is because they are a constant size. It is only by comparing the change in between the relative frequency or height of electron orbits as given by the displacement of the spectral bands on the surface of the sun and here on earth that one can establish the difference between the change in the size of the atoms in absolute space.

I wish all to note I am not trying to disprove relativity. I am trying to disprove the curvature of space. I hold that the expansion and contraction of all material bodies according to relativity occurs in three dimensions. Because space is absolute not relative to motion or gravitational potential. At the same time, I grant you that it is impossible to measure the speed of light as anything other than 299,792 kilometers per second in the small domain. The explanation of relativity most people are familiar with is the speed of light is fixed and constant because space expands and contracts and time slows down or speeds up as the speed of light increases or slows down. This basic feature of relativity appears to be correct, but what most people don't understand about relativity is that Einstein held the speed of light to be variable. These two sentences appear to contradict each other, but the nature of the dichotomy lies in the nature of space.

Einstein held that all space was measured by material bodies, composed of atoms which expanded and contracted according to changes in the speed of light. In short, he held that space was relative in relation to the size of things and dictated by the size of the atoms of which all things are composed. If you hold pure space to be a void this argument breaks down. It becomes an argument that the changes in the size of material things leaves pure space unaltered.

The question becomes how do you measure space? Einstein in explaining relativity used a theoretical set of uniform measuring rods relative to the speed of light. Newton used a concept of absolute lengths in the vacuum of space to measure the distance between the planets and measured the distance between physical objects here on earth in terms of the same absolute space. Einstein used relative tools of measure. Newton used absolute space, absolute lengths. Einstein's measuring tools are all relative to the speed of light. Newton's tools are non-existent but assumed to be the same in all circumstances, under all conditions, whereas Einstein's tools were theoretical and relative to the speed of light.

Let us speculate that the speed of light is three percent higher on the moon. Use relativity's expansion and contraction of material bodies and Newton's space. Thus a six foot tall astronaut is two inches taller on the moon if the speed of light is indeed three percent higher. He will still fit in his space suit because it is three percent larger on the moon, as is his spacecraft. According to Einstein and general relativity the astronaut is the same size. Because there is no physical measurement you could make or take, that would give you the change in his dimensions. The astronaut's height is the same finite distance light would travel in the same finite fraction of one second, regardless of whether this measurement is taken here on earth or on the surface of the moon.

In point of fact there is no known electronic or physical measurement one can take or make that would tell you otherwise. So, the question remains, is the astronaut 6 foot or 6 foot 2 inches tall on the moon? The difference relates to the differences in theory and the effects of gravity across space as we know it. Does gravity work in absolute distances on the atomic level? Does the sum total of the expansion of atoms on the moon, effect the moon's gravitational field? If the answer is yes, then space is not relative. This is a little explored field of gravitational theory. A related question is the planet Mercury five times as dense as the earth or is G higher on that planet? It is speculation in theory. A question related to the growth and reach of theory past its current limits.

Remember Einstein intended general relativity and the curvature of space to be a replacement for Newton's theory of gravity. Because he could not explain gravity as a particle at the time he came up with his theory of the curvature of space, though later in life he attempted to do just that. But before he attempted to do that, Einstein created the intellectual construction of the curvature of space in general relativity.

Einstein explained in detail that in general relativity the presence of matter in space altered the nature of space. Einstein said space was curved depending upon the amount of matter present in nearby space and whether time slowed down or sped up depending upon the amount of matter present in an area of space. It took Einstein twelve years and two tries to come up with an acceptable definition of general relativity for the theory to be accepted by the bending of star light during a solar eclipse. I said two tries. He made an earlier attempt at predicting the bending of star light during an eclipse and was wrong by about half. It was in his second attempt to predict the bending of star light that Einstein came up with an accurate enough figure for his theory of general relativity to be generally accepted. The difference in the two attempts was the idea that the sun's mass slowed down time, which was the final piece of general relativity and how it fell into place.

Albert Einstein was fascinated not only by the contraction of lengths, but also by the fact that when the device contracted in direction of motion it distorted its native geometry, by altering the radius in the direction of motion distorting Pi and quickly came up with the curvature of space. Which held gravity was explainable as motion between two points in space based on the idea that space was curved by distorting Pi. That all things in motion due to the apparent effect of gravity were just following a straight line in curved space. A straight line being the shortest distance between two points.

Einstein went on to elaborate on gravity as motion by comparing gravity and gravitational potential to a rotating disk. He hypothesized that gravitational potential was synonymous to the rate and speed at which the disk spun. The further out you were on the disk the faster the disk spun, the more a set of measuring rods measuring the circumference of the disk would contract in the direction of motion altering Pi. The closer you got to the center of the disk the longer the measuring rods got, because the disk rotated slower. To Einstein gravity's motion was merely matter moving in a straight line. The shortest distance between two points. The distance measured by the fewest relative measuring rods. Motion through curved space was following a hyper-straight path. Straighter than straight as they used to say back in the thirties and forties of the last century.

The ancients used to say heavy object fell because the natural place for heavy objects was the center of the universe, or the center of the earth. Newton changed this paradigm to "matter exerts a force that attracts other matter". Einstein changed the paradigm to everything was motion and things followed hyper-straight lines in curved space. Thus the straightest line was not what appeared straight to us, but rather what was straight in curved space. Einstein said Gravity and the curvature of space were mutually exclusive. Scientists say they prove the curvature of space with atomic time. Not so fast!

What they are proving is Andrei Sakharov's metric elasticity of space, which works on a different set of logical assumptions. Andrei Sakharov was a physicist and mathematician who apparently misunderstood Einstein's basic premises and developed the theory of the metric elasticity of space based on the assumption that the longer rods were out on the edge of the rotating disk and the shorter measuring rods were in toward the center of the rotating disk. Einstein who once made the remark that "Now that mathematicians have gotten ahold of relativity, I hardly recognize it anymore." Failed to catch the error for decades. Nobody else caught it either. So, few people truly understood Einstein's theories, that everybody got lost in a sea of numbers. The proverbial cases of not being able to see the forest for the trees. Andrei Sakharov reserved Einstein's logic and nobody caught it. Everybody is still lost in a sea of numbers. Scientists today can't see the forest because their noses are pressed into the bark of a specific tree.

According to Einstein's original logic Andrei Sakharov's metric elasticity of space meant that large bodies of mass should repel mass, not attract it. Andrei Sakharov was working on an assumption that was the direct opposite of how Einstein thought curvature worked. According to Einstein's logic Sakharov's metric elasticity of space should bend light outward near a large body of mass. Mass should repel mass. It doesn't but nobody noticed the change in theory and the continued to exalt Relativity as the greatest intellectual achievement of the twentieth century. Einstein's curvature is what they claim to prove with atomic time, but now the alteration of Pi that attracts things inward along a hyper long curved path. Let us take a closer look at the mutability of Pi.

The mutability of Pi, flows from the contraction in lengths. That is not the proper explanation for the Michelson and Morley experiment. The proper explanation of the Michelson and Morley experiment is that light rotates as the gravitational field in which it is traveling rotates. There is no contraction in lengths. The displacement of star light during a solar eclipse is always slightly off. It is either too much or too little-hardly ever more than ten percent of the deflection or bending of the star light by the sun's mass. This is due to the irregular rotation of the sun's gravitational field, being faster at the sun's equator than it is at the sun's poles. The rotation of the sun's gravitational field causes the deflection of the star light to be slightly off, depending upon which side of the sun the star falls on. It is a branch of science that presents great promise in the study of the sun's gravitational field, that has been totally ignored and untouched because of the wrong interpretation of the Michelson and Morley experiment.

So, there is no alteration of Pi in the Michelson and Morley experiment or on Einstein's theoretical rotating disk. What are they proving with atomic time, which is based on the alteration and mutability of Pi. The formula for atomic time is a complex fraction involving the energy and mass of the three basic particles neutron, proton and electron, times a second fraction 6Pi over Planck's constant.

$$(M_e/E_e \times E_p/M_p \times E_n/M_n) \, 6Pi/h$$

Wherein E equals the energy of the particle, M equals the mass of the particle, e equals the electron, p equals the proton, n equals the neutron and h is Planck's constant. There are four basic elements 1) The energy of the particles, 2) The mass of the particles, 3) 6pi an approximation of the elliptical orbit of the electrons, 4) Planck's constant a measure of the basic increment of the electron orbital shells. Neither 6Pi or Planck's constant is a specification of the length or diameter of the atom. 6Pi is an approximation of the elliptical nature of the orbits in the electron shells and Planck's constant while a very specific number is variable because each orbital shell of the electron orbits begins on different multiples of Planck's constant. It is a very dynamic equation in which the mass and energy of the basic particles constantly interact with the electron which orbits the nucleus billions of times a second to produce a constant result.

Please note the mass and energy of the electron is grouped opposite the energy and mass of the proton and neutron. The energy of the proton and neutron drive the mass of electrons in its orbit, in conjunction with the energy of the electron driving the electron off the mass of the neutron and proton. The formula is brilliant. One of Einstein's greatest achievements and yes Einstein was very, very good, just not right all the time and he didn't always speak the gospel truth of science despite his apparent status as a deity.

It is not hard to appreciate the exceedingly accurate nature of this equation. Place a cesium clock in an aircraft and fly it around the world. The formula will produce the exact results of the modification of the mass of the particles due to special relativity depending upon if you are flying west to east or east to west. It does this in conjunction with the apparent modification of Pi. Remember 6Pi is an approximate expression of the ellipse of the electron orbits. They don't calculate atomic time for each electron or the number of atoms in cesium clock, but they do calculate the exact number of protons and neutrons in the isotope of cesium that they use in the clock. The formula is very dynamic and exceeding accurate at the same time, but there is a danger in taking it too literally as it is currently interpreted in physics.

Have you ever heard of the sprinter and the snail? The sprinter runs ten times faster than the snail moves. Yet you can construct an overly literal semantical scenario by giving the snail a ten-meter lead in the race the sprinter can never catch and pass the snail. In the time it takes the sprinter to run ten meters, the snail moves one meter. In the time it takes the sprinter to run one meter, the snail moves a tenth of a meter. In the time it takes the sprinter to run a tenth of a meter, the snail moves one centimeter. In the time the sprinter to run one centimeter, the snail moves one millimeter. By the semantics by which you consider the problem the sprinter can never pass the snail. Oblivious you are missing something. The danger in taking the formula for atomic time too literally is similar as again you are missing something. Don't think yourself into a semantical mathematical trap of curvature.

When math doesn't match physical reality, it is pure sophistry. One should read Carl Sagan's book "Cosmos" in the very front of this book he offers a geometric proof that the square root of two is 1.39 not 1.41. He does this using the Pythagorean theorem and right-angle triangle of two right angle sides of exactly one. The Pythagorean theorem states the square of the hypotenuse side is equal to the sum of the squares of the two right angle sides. The square of one is one and one plus one is two. The square root of the hypotenuse side is 1.39. Obliviously the square root of two is 1.41 but are you willing to argue that the

Pythagorean theorem is wrong. Math can be misleading. One can use this mathematical proof offered by Carl Sagan to argue the square root of two is 1.39, but one is arguing sophistry. Math can be sophistry.

Let us consider Doctor Weber's experiment with five atomic clocks which resolved the challenge posed by an Italian scientist of the University of Turin, regarding atomic time. Remember there is no gravity in the curvature of space. Doctor Weber flew five atomic clocks one at a time at an altitude of 40,000 feet for eight hours. To remove the effect of motion in a gravitational field, which he would have encountered if he flew them west to east or east to west, he flew them in circles. Flying one clock a day for eight hours, five days in a row. Each clock gained exactly 30 seconds. An increase in time of exactly .0010416 percent when rounded off. Five clocks over five days produced the exact results. What is missing in the formula that produces a confirmation of curvature? The formula makes no allowance for the expansion of the diameter of the atoms in the clock.

At forty thousand feet the speed of light is .0010416 percent faster in absolute space. In order for Albert Einstein to be correct that you can never measure the speed of light anything other than 299,792 kilometers per second the very atoms have to expand .0010416 percent when rounded off. According to current theory this expansion is absorbed by the 6pi portion of the equation. Pi doesn't change: the diameter of the atom expands. Pi is the ratio of the volume of a sphere to its radius. This is not explicit specified in equation, if it was PI would be cubed. The value of Pi is not squared or cubed as it would be if you interpreted the electron orbits as the function of a circle or the volume of the atom. The atom expands. It doesn't get hyper-straight or hyper-long, it expands. The ellipse lengthens.

Neither Einstein or Sakharov was right in the effect of the formula, the atom does not get hyper-straight or hype-long while staying the same physical diameter. Length is not relative. The volume changes while Pi remains constant as the ratio between the length of the ellipses or the diameter of the atom and its volume. In changing Pi while holding length constant you are apparently changing relative volume. But nobody can describe how that occurs. Nobody can describe how you stack or group these atoms together so that these fundamentally different atoms from a uniform expansion of material bodies interact such that the speed of light always clocks out at 299,792 kilometers per second. I defy anyone to explain how the modification of Pi in atoms effects motion of material bodies through the vacuum of space where there are no atoms to measure it.

There is a basic logical flaw in saying empty space has properties that can be described by changes in matter. If you are saying changes to matter effect space, you are implying the void of space has some properties and is not a void at all. Space is space and matter is matter. By saying the two are interconnected you are taking a leap of faith. You are saying space is not inert without being able to characterize its properties or how they are affected by matter. How are you saying changes to Pi in matter effects its motion through the void of space? You cannot say it works out mathematically so it must be physical law. Do that and you are attributing physical properties to math, similar to the spiritual ones of numerology. Math at its best is the application of abstract concepts to real solid physical entities and forces, not a spiritual exercise. When math does match physical reality, it is pure sophistry

The formula for atomic time is the description of the interplay of the mass and energy of the proton and neutron in direct contrast to the energy and mass of the electron as it plays out across the orbital shells of the atom as defined and demarcated by Planck's constant. You cannot describe the atom as a literal measurement of 6 times Pi, over the value of Planck's constant. That is not physically accurate, it is mathematical nonsense, like the sprinter and the snail. What is physically accurate is to say that the behavior of the atom is a result of the mass and energy of the basic particles of the atom as they interact across the electron orbital shells as defined by quantum numbers or multiples of Planck's constant. You are attempting to put the dynamics of the atom to math, not describing the physical size of the atom. There are no explicit references to the volume or size of the atom in the formula.

What causes the atom to expand or contract in relation to the changes in the speed of light? Raise a cesium clock to 40,000 feet above sea level and its atoms expand 0010416 percent. Descend back to sea level and its atoms contract 0010416 percent. Why? In short, I think the graviton is a polar particle and as its transits space the magnetic fields alternate inhibiting the propagation of an electrically charged particle across that space. This includes the electrical fields of the electron shells of the atom, affecting Planck's constant and the charge of the particles.

The formula for the speed of light is given by the mass energy equivalence formula and is C equals the square root of the energy over its mass of a particle. The energy of the particle is not constant. When an atom changes location into a different gravitational frame of reference with a different speed of light the energy levels of the basic particles change. The electrical fields of the electron shells change and the atom expands or contracts. Andrei Sakharov in coming up with the metric elasticity of space literally came up with the only possible math to describe how it works out mathematically if the theory of mutable Pi was correct. If you assume the idea of changes in Pi you will find his math is always correct. That doesn't make it true. Whether or not you believe Pi changes is an opinion. One reinforced by the mathematical accuracy of the formula. I don't believe Pi changes and if it means

Albert Einstein was wrong in some of his assumptions, so be it and just because Andrei Sakharov made it work out mathematically doesn't mean it is a physically true and accurate description of what happens inside the atom.

Please note I am not trying to disprove relativity. I am arguing against curvature. I accept the idea that things expand and contract according to changes the speed of light. My refusal to accept that Pi changes is driven by the fact that things expand and contract according to changes in the speed of light. That you cannot measure the speed of light as anything other than 299,792 kilometers per second in the small domain. Think about it. The arguments for curvature of space boils down to the idea that the volume of the atom changes while its diameter remains constant because length is relative. If that is the case, how do you combine the atoms together so that your measuring rods contract or expand while the atoms in them maintain the same diameter? I believe in the expansion and contraction of the atoms and material bodies due to changes in the speed of light and that their diameter changes while Pi is constant. I do not believe their volume changes because Pi changes while their diameter is constant. Disprove curvature and you prove a subset of relativity. Just because you can make curvature work as a mathematical argument doesn't mean it is true. I can make 1.39 work as the square root of two, that doesn't make it so. A lot of very smart people are chasing a misleading argument.

Why does the formula for atomic time work? The speed of light is relative, while Planck's constant is a measure of absolute space. As you increment or decrement relative space you must describe these changes in terms of absolute space. You must work with absolute space in equations of relativity.

Relativity makes sense. Atoms expand or contract according to the changes in particles due to changes in the speed of light and because of that you can never locally measure the speed of light as anything other than 299,792 kilometers per second. It is curvature that doesn't make sense. The volume of the atom changes because its diameter changes, not because Pi changes. Relativity works in three-dimensional space with a fixed value for Pi, that is, if you decrement or increment the value of Planck's constant and the energy of the basic particles. In science there is something called Holcomb's razor where the simplest explanation wins. Relativity is a fact and the math that drives it is common sense math.

At this time the changes in Pi are the only explanation of the changes in the size of the atom. The only attempt to account for the change in the size of the atom. PI is assumed to change because no other explanation is given regarding how the atom changes size, when plainly for the mechanics of relativity to work the diameter of the atom must expand or contract. The formula for the speed of light plainly dictates that as the energy of particles changes the speed of light must change.

The only specification for length in the formula is Planck's constant. It is the divisor in the filter to the interplay of the mass and charge of the basic particles in the equation. If the atom expands it alters the relative value of Planck's constant as a measure of the atom. Currently Planck's constant is a fixed value of relative space. You need to decrement or increment the assumed value of Planck's constant in the equation. Because the absolute measurement of Planck's constant changes in relation to the relative size of the atom. To do this you have to apply values of absolute space inside the atom.

I want to say something. Andrei Sakharov's math works with a constant energy of the particle. Do not get too far ahead of yourself in applying math to my physical scenario. Andrei Sakharov's math works because he was in effect, not literally, holding the atom equal to a constant one. It does not work out that way in the real physical world.

Why rework the math of atom time? To make the math easier to work with. Nobody really understands curvature. *It is just a set of numbers and nobody understands curvature except as a set of numbers.* Why not work with changes in the diameter of the atom and a constant Pi, if they make sense? If you can work with common sense numbers and equations to explain relativity, you may be able to break through to the next level of understanding of the physical world. It is stated goal of science to simplify. It would be nice to live up to one's ideals

People think science and math are one and the same thing. They are not. Math cannot be used to develop theory. Math can be used to verify thinking, but ultimately theory must be verified experimentally. Because math can be wrong. Math is a tool for pursuing scientific truths, but math can be wrong or falsely applied. Why does the math of atomic time work as is? Andrei Sakharov created the metric elasticity of space to work out mathematically according to existing theory and it does and always will so long as theory remains constant.

Atomic time is a measure of the vibrations of the atoms. Which is dependent on the electron orbits which are ellipses. The calculation of atomic time is based on the alteration of Pi according to the curvature of space and Lorentz transformation. Einstein held gravity and motion to be one and same thing. As motion distorts time, gravity distorts pi. The calculations work because the electron orbits are ellipses defined as six times pi and Planck's constant. Six times Pi is an approximation. You are affecting the approximation not the inherent value of 6 or Pi. The approximation, the ellipse expands or contracts based on the gravitational frame of reference in which the atom resides.

If you move into an area of gravity with a lower gravitational potential, the ellipse lengthens increasing the volume of the atom. Planck's constant and the energy of the particles change. If you move into a gravitational frame of reference with a higher gravitational potential the ellipse contracts and the volume of the atom decreases. This is what causes the atoms to expand and contract so that the speed of light can never be measured as anything other than 299,792 kilometers per second. If you move at speed you interact with more gravitons as if you were in a higher gravitational potential and the ellipse contracted. This is why motion effects atomic time.

Pi does not change. The diameter of the atom changes. Pi is a ratio of the diameter of circle to its circumference or the radius of a sphere to its volume. If you say Pi changes, you are saying the diameter of the atom is constant. The atom must expand or contract in relation to gravitational potential or else the measuring rods will not expand or contract. Curvature does not work like Einstein thought. Einstein abandoned his ideas of curvature in his last book on relativity but he couldn't explain why atomic time worked, so he didn't denounce it entirely. He was human. He can hardly be criticized for not openly calling his most famous work wrong, especially when other people were convinced it was right and he could not explain why they were wrong. Science is a human endeavor. You should never forget that.

You are in physical reality affecting the ellipse. You are not changing the value of 6 or Pi you are changing the length of the ellipses and the volume of the atom, by altering Planck's constant. I am arguing that pi is a pure ratio, the ratio between the radius of a sphere and the volume of that sphere. This assumes all changes in the atom are due to changes in the energy of the particles and Planck's constant. In quantum mechanics the electron shell is one coherent wave. The total of all the ellipses serves as the diameter of the atom. If you alter Pi in the atom you are arguing for an altered volume of the atom with a fixed diameter. Andrei Sakharov the creator of the metric elasticity of space made the atom expand while assuming a constant diameter. Pi does not change the volume of the atom changes, but so does its diameter. Space and numbers are absolute. The alteration produced in the electron orbits is because of Planck's constant modification changing the value of the electron orbits. The cumulative total of the ellipses is the diameter of the atom. He was describing the ellipse of the electron orbits, which expand and contract as the charge of the particles changes and Planck's constant changes.

Pi does not change in relation to gravitational potential. The radius and volume of the atom in absolute space changes, not Pi. The radius and volume of the atom has to change or the measuring rods will not expand and contract in proportion to the changes in the speed of light. There are no changes to Pi just as there is no contraction in length in relation to motion. The proper explanation of the Michelson and Morley experiment is light rotates as the gravitational field in which it is traveling rotates, not a contraction in lengths. Length is not relative to motion, in the Michelson and Morley experiment or on Einstein's theoretical rotating disk. Pi is a pure numerical ratio and is not relative to gravitational potential.

Einstein intended the curvature of space to be a replacement for gravity. The logic which he used to do this has been completely discredited. There is no such thing as the curvature of space, only gravity. Pi is unaltered because space is absolute. I repeat you are not altering Pi you are altering the approximate value of the ellipses which are the electron orbits, based on Planck's constant. Length is not relative: if the ellipse expands, it expands. All electromagnetic effects are relative, frequency and wave length change according to relativity, but space is still absolute. Gravity attraction works in terms of the absolute space between two masses through the polar graviton, not the alteration of Pi.

Of Science and God

Scientists must face up to the possibility that Andrei Sakharov came up with a mathematical system which is basically correct despite the fact that it has no relevance in physical reality or to the conditions inside the atom. They must be willing to abandon their math and step off into the uncertainty of new theory. The math may not work out. A new formula for atomic time may be needed. Scientist are currently hostage to the math and working on the math of a theory is not necessarily the same thing as developing new theory.

The idea that the charge of the elementary particles changes according to the gravitational potential of the area in space where they are measured, once proven invalidates the argument that atomic time proves the metric elasticity of space due to changes in Pi. You cannot have both a variable value for Pi and variable charge for the elementary particles. One is right and the other is wrong. Einstein was right Gravity and curvature are mutually exclusive one is right and the other one is wrong.

Can you calculate atomic time with a fixed value for Pi? Logic is a guide but not a guarantee. Meanwhile, I will continue to describe what I see as the changes in the physical characteristics of the atom to fuel the scientific debate. But at some point, in the future, someone may have to take a massive leap into the unknown.

For all the arguing and assertions so far in this book, I have not disproved relativity, I am not trying to disprove relativity. I am trying to disprove curvature. I am attempting to resolve the differences between gravity and relativity, by dispensing with the curvature of space. Space is absolute not curved with wormholes and all those so-called advanced ideas. I have read about in books, about how the curvature of space effects the universe.

Some smart and clever physicist or mathematician might make the argument that gravity creates the exact same effects and changes in space that metric elastic space does. There is a lot of room for debate in such an arguement or assertion. The reason I don't believe it, is the fact that gravity bends light inward not outward as Einstein explained in his original explanation of the curvature of space. If the metric elasticity of space was true, the curvature of space would bend light out near a very strong gravitational source. Albert Einstein was wrong about how curvature worked. I think he was wrong that space was curved, period.

I want to be very clear about this. You can argue the change in the charge of the atomic particles and Planck's constant has the effect of shrinking or expanding space. That light follows the longer slower path near a large mass. This argument is supported by the fact wave length and frequency are relative and counterposed by the argument there is no distortion of Pi. The difference is if you believe gravity attracts things and not the metric elasticity of space controls motion through curved space through the distortion Pi.

The variable speed of light and Time

"The atom is like a clock"
a saying of Albert Einstein

It is not for nothing that Stephen Hawking named his book. "The History of Time". The accuracy of Einstein's formula for the change in time due to motion and changes in gravitational potential are the principal facts upon which scientific belief in relativity's curvature of space rests, but the metric elasticity of space has been plainly demonstrated to be wrong by the inward bending of starlight by strong gravitational fields.

Albert Einstein observed the atom is like a clock and provided the formula by which atomic time is measured. It is a complex fraction involving the mass and electrical energy of the elementary particle of the atom proton, neutron and electron, multiplied by a second fraction involving Planck's constant and six times Pi. The reasoning is "the formula works so the metric elasticity of space must be true." Scientists are failing to account for the differences between Einstein's original theories and current theory. Let us tackle the issue of atomic time.

If there is no metric elasticity of space and space is absolute, the speed of light is variable and scientists must use a variable value for the speed of light in absolute space when calculating scientific effects and applying the formulas of physics. Most people do not realize the speed of light is variable. As a result, the periodic rate of time is variable as well. Remember the speed of light, time and space must contract and expand at the same rate for relativity to work. Time is not a dimension. Albert Einstein originally described time as a coordinate of observation. Over time as the theory of relativity developed, there arose a concept of space-time as a dimension. This is extremely misleading.

We live in shells or layears of time. An electronic clock taken to the second floor of a house or apartment building and left there will in five to six years will gain one second on an electronic clock on the ground floor. This is directly due to the fact that the speed of light is slightly faster on the second floor of a house than it is on the first floor. One can say we live in shells or layears of time as you rise off the earth. Think of an infinite onion with incredibly thin layers or shells of time. These layears of time are different rates in the periodic rate of time, which is faster or slower depending upon the speed of light in that layer.

Obviously, people don't have five or six years to compare two electronic clocks on different floors of their house and the experiment assumes the power supply is not interrupted. A faster way to prove this concept is to up the interval in between the two layears or shells of time to one thousand feet in height. The Eiffel Tower or the observation desk of the Empire State Building both are about one thousand feet in height. Take two electronic clocks and place one on the ground floor of the Empire State building and place the other in the hall-way to the observation deck in that building and they will run at different rates. In about 18 days the clock in the hall-way to the observation deck will be one second ahead of the clock on the ground floor. You can do the same with two clocks and the Eiffel tower. The rate will be slower for the Eiffel tower, because the restaurant at the top of the Eiffel tower is not quite as high up as the observation deck of the Empire State Building. Time is not a dimension it is the periodic rate of change.

Time travel makes for great science fiction but it is practically and physically impossible. Because time is not a dimension, it is the variable periodic rate of change. Geological time is a different form of time than common human time. Once you understand this concept of differences in the periodic rate of time, you understand the impossibility of time travel. Time is a variable coordinate of observation. As you change the different shells of time you do not travel to a different dimension. A waiter who spends twelve hours a day serving meals at the restaurant at the top of the Eiffel tower ages almost a second more in a month, than the vendor selling hot dogs on the ground beneath the tower. Say the waiter in the restaurant at the top of the Eiffel tower and vendor who sells hot dogs on the ground are brothers and live in the same house. If they sleep in adjoining rooms on the same floor of the house, they literally work in different areas of time, but sleep in the same area of time. Think about this concept of time and take it to heart, before you try to understand the relative concept of spacetime.

When it comes to time, it is a question of what you are measuring, where you are measuring it and how far removed you are from the event you are measuring. Time on the surface of the earth is different from time on the surface of the moon. If a star novates and explodes in space it will be observed at different times in different places in the galaxy. The speed of light is slower at the center of the galaxy than it is on the edges of the galaxy or in the space between the spiral arms of the galaxy. Because the ratio of mass to space and G is different in all those areas.

If you want to understand spacetime consider the following proposition. A white dwarf star novates on the edge of one of the spiral arms of a galaxy. The white dwarf is located toward the center of the galaxy. Where the arms of the galaxy are thick and relatively close to together. If you further say the distance between the spiral arms of the galaxy are the same as the thickness of the arms at that point in the galaxy. There are two points of observation. One on the opposite side of the spiral arm. The second on the next spiral arm over in the galaxy. The white dwarf is literally the same distance in absolute space from both points of observation. When the white dwarf novates, the explosion will be observed on the different spiral arm before it is observed on the opposite side of the same spiral arm. Because G, the ratio of mass to space, is different between the spiral arms, then it is through the body of the spiral arm. The stars on the spiral arm increase G as you travel through the body of the spiral arm. While there are no stars between the spiral arms dropping G. That is spacetime, the speed of radiation through space. Astronomer understand the concept of time as follows: If you are looking a star four light years away you are looking back in time, four years.

A simple concept, but space time is more complex than that. Stephan Hawking theorized that black holes can radiate energy and I believe him. This radiation crawls off the event horizon of the black hole at speeds of less than a hundred miles per hour in absolute space. But in a month's time that same radiation is traveling hundreds of thousands of miles in absolute space in a second. The speed of light between galaxies is not 186,000 thousand miles per second in absolute space. It is maybe as high as a million miles per second in absolute space.

The science to properly calculate these afore mentioned effects simply does not exist in science. Because everybody thinks the speed of light is fixed and constant at 186,000 miles per second or 299,792 kilometer per second. Space is thought to expand and contract at the same rate as the speed of light. They do not work in absolute and relative space in astronomy or in physics for that matter. Space is supposedly metrical elastic and the metric elasticity of space is confirmed, by atomic time. I submit to you the speed of light is not fixed and constant. That the idea the speed of light is fixed and constant is an illusion that science can no longer afford. Please however note if you were on a flying sauce at 95 percent of the speed of light escaping that nova of our supposed white dwarf it would take a similar amount of time to travel to both points of observation, according to a clock on the flying saucer. Because the periodic rate of time would change with G and speed at which you were traveling. Relativity is an ass kicker. Space may be absolute, but don't underestimate what Albert Einstein wrought.

The math and knowledge to calculate the difference between the relative travel time it would take to travel between the spiral arms versus the relative time it would take to travel through of one spiral arm does not exist. Because they have never measured the differences in the periodic rate of time as they traveled to the moon and Mars or beyond. Very careful calculations of the differences in the periodic rate of time as they travel to Mars or beyond would give them a better idea of how much of G here on earth is background radiation and how much is due to the gravity of the earth and moon.

The periodic rate of time varies directly with the speed of light. Time is without a doubt relative. The formula for the speed of light is known in science but not generally acknowledged. The speed of light is given by the mass energy equivalence formula and is C equals the square root of the energy over the mass of particle. Taken literally the mass/energy equivalence formula dictates that the energy of a particle must change when it enters an area of space with a different local speed of light. This is a relationship of the energy of a particle and the field forces of space. I call the electro-magnetic index of space.

$$C = \text{the square root of } (E/M)$$

This is a relation between the particle and space. I believe the graviton is a polar particle with positive and negative poles that energizes the electrical resistance of space. The graviton is far more massive than the electron and the atom does not emit gravitons. It interacts the gravitons that already exist in space. As the polar gravitons travel through space these negative and positive fields alternate, inhibiting the flow of electrical energy across that space. Even if that space is the distance between the electrons and the nucleus of the atom. The more gravitons the greater the resistance of space to electrical energy, lowering the effective energy of the particle. The less gravitons the higher the effective energy of the particle. This directly translates into an increase or decrease in the speed of light and the periodic rate of time.

There are two different types of atomic clocks, one chemical (ammonia) and the other electro-magnetic (cesium). The ammonia clock works in a frequency of slightly over 23,000 hertz, while the Cesium clock works in a frequency of over nine billion hertz. The ammonia clock is not used in experimentation because the changes to the time of an ammonia clock due to motion cannot be calculated mathematically. Whereas the changes in relativistic time in a cesium clock can be exactly calculated mathematically.

The seconds lost or gained should be the same in any experiment involving relativity, independent of the clock used, so the prejudice for one clock over the other is purely an intellectual prejudice. I want to say something about scientific experiments with time. They should repeat their experiments with both chemical clock and a mechanical clock, based on the mechanical energy of a spring. Why? Because gravitational meters are based on very sensitive springs. If springs conform to electromagnetic effects of relativity, you may not find a difference in between the measurement of G on the moon and the different planets. This means you would have to very carefully explore differences in time as a measure of the changes in G and the speed of light on the different bodies in space. It should be noted, the ammonia is a measure of chemical time and that is the same time as the human body. If chemical time is different than atomic time in motion, then human time in space travel is at the rate of chemical time, not atomic time.

Einstein's formula for atomic time involves the energy of the elementary particles. In this fraction the mass of the proton and neutron is grouped together with the energy of an electron and the mass of electron is grouped together with the energy of the proton and neutron. Einstein was a great believer in the three forces atom and was asserting that the mass of the elementary particles and electrical repulsion for each other controlled the speed of the electron orbits-that the protons and neutrons of the nucleus and the electrons push each other in opposite directions. The speed and height of electron orbits is directly attributable to Newton's laws of force and electrical repulsion of elementary particles of the atom for each other.

Again, the formula by which atomic time is measured is a complex fraction involving the mass and electrical energy of the elementary particle of the atom proton, neutron and electron, multiplied by a second fraction involving six times Pi and Planck's constant.

$$(E_p/M_p \times E_n/M_n \times M_e/E_e) \times 6Pi/h$$

Wherein E is the energy of the particle, M is the mass of the particle, "$_p$" stands for the proton, "$_N$" stands for the neutron, "$_E$" stands for the electron and h is Planck's constant. According to current theory, atomic time is calculated based on alterations to Pi according to metric elasticity of space and Andrei Sakharov's math for atomic time.

Physicists are literally contending that the presence of matter alters the properties of space. There is an inherent logic flaw in measuring changes in the properties of space, by measuring the changes in matter. Matter is matter and space is space. It does not necessarily follow that changes to matter effect the void of space. It takes a leap of faith to say that changing the properties of matter changes the properties of space.

Einstein insisted that gravity and curvature were mutually exclusive. I say gravity effects the electrical properties of space, regardless of whether that space is the void of the empty space of the cosmos or the empty space inside the atom. Again, the more gravitons that transit the atom in given period of time the greater the resistance to the propagation of electrical energy across that space, changing the effective charge of the elementary particles: electron, proton, neutron and photon. Gravity through the polar graviton sets the electro-magnetic index of space.

This occurs in a manner similar to the way the electron bonds of a transparent substance set the refractive index of a transparent substance, controlling the speed of light through that substance. As the electromagnetic index of space goes up the effective charge of elementary particles and the speed of light goes down. As the electromagnetic index of space goes down the effective charge of the elementary particles and the speed of light goes up. Remember that the lower the electromagnetic index of space the higher the speed of light. The higher the electromagnetic index of space the lower the speed of light. Is this speculative theory?

Doctor Weber of the University of Maryland tested the theory of the metric elasticity of space by flying five atomic clocks at an attitude of 40,000 feet for eight hours. He flew them in circles to remove any changes in time due to motion in a gravitational field. All five clocks gained the exact same amount of time, 30 seconds. The experiment confirmed the alterations of PI according metrical elasticity of space. This sets a very high bar to overcome in contenting that space is absolute not curved. You must explain why the formula works when Pi is constant. This change of 30 seconds in 8 hours corresponds to one in 960 parts or .0010416 percent when rounded off.

The formula for atomic time works when you calculate the change in Pi at altitude according to the metric elasticity of space. If there is no curvature of space, no alteration of PI, then what changes? Break down the formula for atomic time: for the sake of argument we are ruling out changes to Pi. That leaves the mass and charge of the particles and Planck's constant. Mass has been proven to be relative in relation to motion in a gravitational field, but Professor Weber removed that factor in the experiment by flying the clocks in circles. That leaves the charge of the elementary particles and Planck's constant.

First, let us discuss the charge or energy of the elementary particles. Another way to look at the problem of atomic time is: the effective charge of the elementary particles of the atom increase or decrease in gravitational fields of varying strength. Remember the greater number of gravitons pass through any given area in space in a given period of time the greater the resistance to the propagation of electrical energy across that space, decreasing the effective energy of the basic particles.

The lower the number of gravitons in an area of space, the higher the effective energy of the electron, proton, neutron and photon. The higher the number of gravitons in an area of space, the lower the effective energy of the electron, proton, neutron and photon. The effect of motion upon time can be said to be due to the fact that the faster you go the more gravitons you interact with.

Remember the mass and energy of a particle are linked by the mass energy equivalence formula. $E=MC^2$. Which also explains of the speed of light when stated as the speed of light equals the square root of energy over mass of the elementary particles.

$$C = \text{square root of } (E/M)$$

This is the relationship between the mass of the particles and their energy, as given by the square root of the energy of the elementary particles over their mass. Applied literally it dictates changes in the energy of a particle as it travels through areas of space with a different local speed of light. Stated differently it is the effective charge of the basic particles of the atom across the empty space of the electron orbits. It is controlled by the number of gravitons that transit the mostly empty space of the atom in any given period of time. This is the electromagnetic index of space. I got that name from the refractive index which measures the speed of light through a substance.

There is evidence for a change of the effective energy of the elementary particles in science. This evidence consists of the difference that is manifest between the old analog electronic and chemically produced images from space. The old analog television cameras work differently from the newer digital cameras. This effect was pronounced in the older analog television broadcasts from space and the moon, when compared to the pictures taken in space or on the moon. There was a slight discoloration in the old analog television pictures of the astronauts on the moon, which is not present in the photographs or the newer digital television broadcasts. The electronic images were tainted, discolored, like there was a filter on the lens.

In the analog broadcast of images from the moon and/or space the increased energy of the photon is captured and that increased energy is broadcast to earth where particles have a different level of energy and the analog video display equipment experienced a different level of energy. This produced a discoloration in the image. A pronounced violet shift or frequency shift of the images. The formula for frequency of light is energy of the particle over Planck's constant equals frequency.

$$E/h = V$$

A different energy produces a different color according to the formula. Due to this violet shift white was no longer white even as it should have been with black and white cameras. This difference was not manifest in the original Apollo 11 television broadcasts from the moon. Because the first nuclear reactor they put on the moon suffered a ten percent power loss from its measured performance here on earth. In that broadcast white was white. Neil Armstrong's space suit was a bright and clear white in the first broadcast from the surface of the moon, but the lighting was bad.

They then fixed the problem with the power supply by the second moon mission the lighting improved but the spacesuits were not white but rather violet tinted. Due to the frequency shift, the astronauts' suits were not pure white as they should have been even with a black and white camera. I thought it was my color television set, but Walter Cronkite's shirt was pure white. So, it wasn't the television set it was the broadcast from the moon.

At the time, I wrote this off to the idea they were using black and white cameras, but over the years it occurred to me that in the Apollo 11 TV broadcast the astronaut suits were pure white, because there was a change in the electrical energy levels to the cameras of Apollo 11 from the first nuclear reactor power loss. The astronaut's suits were pure white in that broadcast from the moon. What changed in between the original broadcasts and subsequent broadcasts?

All this may sound like anecdotal evidence but visual evidence is considered empirical evidence and if you don't believe me you can go to the old video tape library and look at the original broadcasts. A similar discoloration was also manifest in the images of the clouds as captured by the old analog weather satellites. The clouds were a distinctive gray again as if there was a filter on the lens. In the pictures taken from the moon the earth's clouds were pure white and the same is true of them in the Imax film taken from space shuttle. The violet shift or darkening of the television images from the moon and space was slight but distinctive, especially when looking at something white. Again, I repeat, white should be white, even when viewed with a black and white camera.

One must account for other factors such as lighting. The photographs the astronauts took on the moon, however, were perfect and had vivid color so it couldn't have been the lighting. The sun is after all a massive arc light. Was it a shadow? Shadows on the moon are pitch black because on the moon there is no atmosphere to disperse the light, like the atmosphere does here on earth. One must conclude that conditions on the moon and space were such that the cameras did not behave electronically the same as they did here when constructed here on earth. What changed?

The problems with the old analog camera's do not manifest themselves in the newer digital cameras. Astronomers have stated there is no difference in between the images from the Hubble telescope and earth-based telescopes. Additionally, the Chinese government has landed lunar rovers with the newer digital cameras and they don't show a frequency shift. Indeed, the images from those cameras appear to be much the same as the photographs and the naked eye. The images are very earth like and manifest no frequency shift. What is different in construction of the old analog cameras and the newer digital cameras?

You can't just say NASA's cameras were defective. They were state of the art the best in technology at the time. There is something inherently different in the old analog broadcasts and the newer digital images and broadcasts from space. A difference not manifest in comparisons of analog and digital images here on earth.

The nature of the difference between the television analog and digital broadcasts lies in the construction of photodiodes. They have what is known as a reverse bias. They transform the violet shift of one diode with a red shift in the second diode leading to a null effect. This is an extremely well-documented effect and occurs in photodiodes which are classified as junction diodes. The term is derived from diodes that are composed of two diodes which are connected together through an interface. The effect can be compared to the double refraction of light through two prisms. The first prism breaks white light down into the color spectrum. The second prism turns the color spectrum into white light again. This reverse bias effect when applied to the Hubble telescope or digital cameras on the moon produces an earth-like image.

The distortion of the image in analog broadcast from space is due to the increased energy of the photon. I repeat, the formula for frequency is:

$$E/h = V$$

Wherein: E is the energy of the photon, h is Planck's constant and V is Frequency

This would explain the increased color frequency of analog broadcasts from space. It also explains why the energy of the photon registers differently in the same analog camera at sea level and in geosynchronous orbit. This is manifest in the frequency shift of an analog camera in geosynchronous orbit.

Normally the wave length of the radiation decreases with altitude as you climb out of a gravitational well, but the nature of electronics and the increase in wave length inverts this decrease in frequency decrease into a frequency increase. In the previous chapter I maintained and asserted that Planck's constant and the height of the electron orbits or frequency of the electron shell increase or decrease with changes in the speed of light. This leads to a direct contradiction in Physics theory. How can frequency increase when according to changes in Planck's constant frequency should decrease? I cannot explain that contradiction at this time.

There is a formula in Astronomy that goes directly to the lengthening of the wave length of radiation as you climb out of a gravitational well. It is:

$$GMD/C^2R$$

Wherein G is the Gravitational constant, M is the mass of a celestial body, D is the distortion of the wave length, C^2 is the speed of light squared and R is the radius of the celestial body.

This is thought to be because of the pull of gravity lengthens the wave length of radiation as it climbs out of a gravitational well. This makes sense if Planck's constant increases or decreases with the speed of light in absolute space. In addition, according to the electromagnetic index of space the energy of the particle increases as the speed of light in absolute space increases. As a general rule in science the longer the wave length of the radiation, the lower the energy of the radiation. This is one hell of a contradiction and threatens my theory of the electromagnetic index of space and I cannot resolve these differences at this time.

I grew up thinking Albert Einstein was guilty of sorcery and in relativity cast a spell. Pi was still Pi, no matter where you were in the universe. Albert was brilliant and the effects of relativity are more inclusive than I understood reading Albert Einstein at age ten and twelve. I think the effects of the change in the electromagnetic index in space increase or decrease in Planck's constant in direct proportion to changes in the speed of light and that the energy of the particle is translated directly in an increased wave length in absolute space of that radiation. All of these effects act according to the dictates of relativity and conform to the following formulas.

First, Planck's constant increases by the following formula:

$$(h_1/h_0) = (C_1/C_0)$$

Two the wave length of the radiation increases according the following formula:

$$W_0(_1C/_0C) = W_1$$

The increased energy of the particle plays out over an increased distance in absolute space. The relationship between wave length and frequency as the electromagnetic index of space is counter intuitive. As a photon's energy increases with an increase in the speed of light the wave length lengthens. It does this despite the fact the frequency increases as well. The relationship between wave length and frequency as the electromagnetic index of space changes is given by the following series of formulas:

$$W_0(_1C/_0C) = W_1$$

$$E_0(_1C^2/_0C^2) = E_1,$$

$$E_1/h_1 = V_1,$$

$$V_1 W_1 = C_1$$

Wherein "$_0$" is the electromagnetic index at the source of the radiation, "$_1$" is the electromagnetic index at the place of observation, h is Planck's constant, C is the speed of light, C^2 is the speed of light square, W is the wave length, V is frequency and E is energy. The second set of formula can be confirmed by comparing them to the figures of the formula for the redshift of light as you climb out of a gravitational well from astronomy given previously in this chapter.

How does one go about proving the electromagnetic index of space? Putting a laser in a high-altitude balloon offers a method. As is the projection of a laser beam up or down a significant altitude of a thousand feet or more. You can also possibly measure a laser beam projected up into a high-altitude balloon. At forty thousand feet the atom is .0010416 percent larger. The change in wave length should be proportional to that value. Making it difficult to impossible to measure. The frequency is another matter. Frequency would increase by the formula:

$$E_1 / h_1 = V_1$$

Wherein: E equals energy, h is Planck's constant and V is frequency of the laser at the altitude it is when it is emitting photons. The energy of the laser is the result of the increase energy of the electrons stimulating the laser's emissions and the energy of the particle of the atoms emitting the photons interacting across an increased distance within the atom due to the expansion of Planck's constant. The mass of the photon should go down because of the higher speed of light. The formula for the mass of the photon being:

$$M = V_1 h_1 / _1 C^2$$

Wherein: M is the mass of the photon, V is the adjusted frequency, h is the adjusted value for Planck's constant and C^2 is the speed of light squared at that altitude.

The changes in wave length should be the same for the associated frequency.

$$W_0(_1C/_0C) = W_1$$

The resulting new frequency when multiplied by the change wave length equal the increased speed of light in absolute space.

$$V_1 W_1 = {_1}C$$

The change in wave length of laser light would be very difficult to measure. Think of this way. Your measuring equipment will expand with the changes in speed of light. While the relative frequency in terms of the periodic rate of time decreases. The decreased frequency times the expanded wave length equals the variable speed of light in absolute space. The expansion of the relative wave length will match the expansion of your measuring equipment. Plainly you have to experiment and take frequency and wave length measurements. Both with lasers projected up and lifted up in altitude. Please note if frequency decreases when you project a laser beam up. Frequency should increase when you project a laser beam down in altitude. This effect may be measurable in the elevator shafts of skyscrapers.

I said in the foreword to the book I wanted to disprove curvature while proving a subset of relativity. Einstein intended the curvature of space to be a replacement for gravity. In do so he created the curvature of space or relative space. As I pointed out repeatedly in earlier chapters relativity is little understood and the idea the speed of light is fixed and constant is illusionary and leads to a misunderstanding of Physical law.

The idea that space expands and contracts with changes in the speed of light is misleading and may be mathematically unsupportable in consideration of Gravity. Just like the idea the speed of light is fixed and constant is misleading. Space is absolute and to understand relativity one must be able to work in both absolute and relative space.

The average reader thinks time is a dimension. They eat up and love stories of time travel. In the science fiction you make the jump to hyperspace by folding space and time. Complete bullshit, the average reader should not feel bad however, because the average scientist doesn't understand the difference between relative and absolute space. Science is full of examples of curvature and changes in curvature due changes in time. Space is thought to be relative not absolute because of the accuracy of atomic time. What the average scientist does not understand is that space is absolute, while the size of all material or physical bodies is relative. There is a difference in between the size of material or physical bodies and absolute space. Once these differences are understood, the expansion and contraction of the atoms in physical bodies and the relative nature of time become Physical law and Einstein theories with some modification become Physical Law like Newton's laws of force and motion and Newton's principles of Gravity.

The only thing that standing in the way of this is curvature of space. Nobody understands curvature except as a set of numbers. Scientific belief in curvature and the metric elasticity of space is the direct result of Andrei Sakharov's work in atomic time. The accuracy of Sakharov work in time according to Einstein formula for atomic stands up to every test ever devised for it. Again, the formula for atomic time is:

$$(E_p/M_p \times E_n/M_n \times M_e/E_e) \times 6Pi/h$$

They have put atomic clocks and flown them around the world the formula checks out exactly on the assumption Pi is mutable. The

Calculations for the satellites used in the Global positioning system. Is based on this formula and the mutability of Pi. Atomic time and the mutability of Pi has stood up to every time ever posed by a physicist and is based on curvature. To kill off curvature, hold Pi and numbers to be absolute standards and restore gravity to its rightful place in physics and astronomy one has come up with a way of calculating atomic time, with a fixed and constant value of Pi. I do not understand the relationship between G, the speed light as given in Andrei Sakharov's metric elasticity because I never took Trigonometry, but I am going to give it a shot.

I submit the following modified formula for atomic time to you:

$$(_1Ep/Mp \times {}_1En/Mn \times Me/{}_1Ee) \times 6Pi/{}_1h$$

Again, the notation of "$_1$" is a notation reflecting the electromagnetic index in the place of observation. If the notated formula is one part per 960 parts different than the original un-notated formula at sea level for atomic time at forty thousand feet as in Weber's experiment. I have put a stake in the heart of the demon of curvature and killed it as surely as you kill a Vampire.

I want you to note the speed of light and atomic time is changing here on earth. To understand this proposition, read the next chapter on Cosmology. Where I explain how and why our image of the universe is wrong. G is dropping and C is increasing here on earth due to our fall towards the great attractor.

Cosmology

"All was darkness, then there was light"
From the Bible describing creation and the
intellectual origins of the Big Bang theory.

The question is: is our current picture and perception of the universe, correct? According to current thought the universe is one coherent entity and shaped like a massive donut due to the metric elasticity of space. The curvature of space effects the size and dimensions of the universe. As we have noted if space was curved or metric elastic, starlight would bend outward around intense gravitational fields. It does not because empty space or space in a vacuum is not curved. So, should we dispense with all relativistic cosmological assumptions? If space is not metrically elastic, not curved, what is it then? One is left with the conclusion that space is three dimensional with time simply a relative coordinate of observation as Einstein originally described it and not a dimension at all. The question remains; What is shape and nature of the Universe?

All current cosmological assumptions are based on the curvature of space and the big bang theory. The big bang theory has its origins in theology. A Jewish theologian posited that at the moment of creation all matter in the Universe must have been in one place. At a scientific conference at the Vatican hosted by Pope John Paul II, John Paul told the assembled scientists, not to mess with the big bang theory because the Catholic Church considered it to be moment of creation. We must separate science from religious dogma. In doing so we find that in this case, religion and science share the same dogmatic concept, that there was a specific act of creation for the entire Universe. Let us examine the possibility that the universe is both infinite and eternal, with no beginning and no end, in size or time.

The background microwave radiation of space is considered to be the major proof of big bang theory. In the 1950s a team of three scientists, including Professor Gamow (who was Christian) published the theory that if the big bang happened space would be full of microwave radiation. This was light from the original big bang frequency down shifted. In 1962 a team of scientists from Bell labs using high attitude balloons discovered space was full of microwave radiation and attributed it to this theory. The big bang theory has been a fixture in science ever since. The case for the big bang theory however is not so clear.

The scientists who posited this theory of frequency down shifted light, assumed the universe had positive curvature, that the curvature of the universe was spherical in nature according to Einstein's theories. Positive curvature means any object or radiation traveling outward in the Universe would eventually bend back on itself. This would account for the microwave radiation. It is the light from the Big Bang frequency shifted and bent back into our area of the universe by the Universe's positive curvature.

However, subsequent studies of the Universe and its supposed curvature strongly suggest negative or a saddle shaped curvature. Meaning any radiation or body traveling outward would continue to travel outward for all eternity. Not bend back in at all. So, no matter if you believe in a curved universe or a three dimensional one, the background microwave radiation of space cannot possibly be light from the Big Bang frequency down shifted, because that requires positive curvature. Astronomers are convinced the universe has negative of saddle shape curvature. According to current theory the light from the big bang would never bend back inwards instead it would radiate outwards forever. Astronomers have never addressed or explained this contradiction in theory, but instead continue to state the universe was born in one massive explosion and has been expanding ever since.

If the background microwave radiation of space is not frequency down shifted light from the big bang, what is this radiation? Without reference to General Relativity Stephen Hawking proved by logical construct or one of Einstein's thought experiments, that black holes can radiate energy. Is the background radiation of space Hawking's radiation from the black holes? Black holes surround us in the Universe as a whole and if Stephen Hawking was wrong there is no known force or effect to explain this radiation. But I do know that the background microwave radiation of space cannot possibly be light from the big bang frequency down shifted, because that argument requires the positive curvature of space.

Another observation which supports the idea of a three dimensional and infinite universe with no beginning or no end, in size or time, is the discovery of extremely distant galaxies by an astronomer who aimed the Hubble telescope at an empty and dark space in the cosmos. He photographed it numerous times and then stacked 27 photographic plates on top of each other, only to discover nearly a dozen new galaxies. The most distant celestial objects ever observed. Two of these galaxies were visibly red shifted as if receding from the earth at great speed. According to the conventional wisdom red shifts being the result of motion. The rest of them were standard white light suggesting their position in relation to the earth was relatively static. By current logic in astronomy if these galaxies were expelled outward from the big bang and they all should have been red shifted. They were not! In terms of the big bang can you explain, why the most distant galaxies ever observed, are not receding from us at great speed? If one repeats the observation experiment with other dark and empty areas of space, would one find additional galaxies? Try it and see.

Stephen Hawking has theorized small black holes in an area of space with a lower value for G can explode. Invert Hawking's argument and you have an argument that an intense gravitation source or object in space devoid some independent sources of gravitational radiation might explode. Inversions happen in physics. If a black hole becomes massive singularity within a massive void and external background sources for gravitational radiation dramatically decreased would that black hole explode?

There may be evidence that black holes explode, in current astronomical observations. It is contained in the famous Hubble telescope's photograph of the massive dust cloud spanning millions of light years, known as the pillars of creation. If this dust cloud is the result of the explosion of a black hole it still should be expanding. It is important to once again observe this dust cloud and look for signs of expansion. This may be difficult. Given the size and scale of the dust cloud it would take decades or centuries for the features of the dust cloud to change enough to be noticeable to us here on earth.

The universe is both infinite and eternal and has no beginning and no end, in size or time. There was no specific act of creation. It has always been here and always will be. There is no expansion of the universe from the big bang. It is three dimensional, with galaxies moving towards other areas with a greater ratio of mass to space. While other areas are expanding from the explosion of massive black holes. Which leads to formation of new galaxies in motion.

This explains the observation of how galaxies moving in different directions collide. If there was only one big bang the distance between the debris would constantly be expanding, as debris expanded outward into the void. The galaxies and the distance between them would be constantly expanding. Yet there are a countless number of examples of galaxies colliding. The astronomical photographs are compelling to look at. Colliding galaxies are impossible if the universe was formed from one massive explosion a few billion years ago. If the entire universe was formed from one massive explosion, the distance between the galaxies would be constantly expanding in an outward plume like the shards of a grenade. The shards of a grenade do not collide in flight, unless another grenade exploded nearby at exactly the same time. There has to have been subsequent explosions or collapses of portions of the universe. There was not one big bang but several smaller explosions.

Another observation that seems to support the big bang theory is the red shift of Starlight, that increases as distance from the Milky Way increases. This red shift is attributed to motion and is used to map the universe. Quasars or extremely red shifted cosmic formations are thought to be the most distant and fastest moving things in the universe. These quasars are cosmic entities that produce massive volume of radio emissions and relatively little light. Quasars are thought to be on the very edges of the Universe. This is exclusively due to their red shift. Let us examine the proposition that this red shift in starlight is due to differences in G and the speed of light in other parts of the universe. This argument is predicated on the idea that the universe is an aggregate. With no universal constants, no constant speed of light or fixed value for G.

The amount of quanta in a photon or mass of that photon is fixed at its creation. When a photon is created in an area of the universe with a lower speed of light due to a higher value for G and that photon transits an area of space with a higher speed of light and lower value for G, the observed frequency and relative wave length of that photon changes.

Let us deal with the redshift of starlight according to the formula:

$$_0W(_1C/_0C) = {_1W}$$

Astronomer assume the changes in wave length, the redshift of star light, is due to motion. When it could be easily changes in the corresponding wave length of a photon due to changes in G and the variable speed of light. These changes are due to light transiting area of space with a different speed of light. Using this formula one can figure out the differences in the speed of different area of universe based on the redshift. It will change the map of the universe. If one accepts that area of the universe with different gravitational potential have different speed of light. Then one is forced to conclude that photons originating in area of the universe with a different speed of light have different wave length when observed in areas of the universe with a different speed of light. The conclusion is inescapable.

You can check the general veracity of this formula as producing differences in the wave length of observed starlight according to difference in the speed of light by using the formula for displacement of the spectral bands of the hydrogen atom according to the following formula.

$$(V_0/V_1) = (C_0/C_1)$$

Wherein V_0 is the observed spectral bands of a hydrogen atom in the sun or other celestial body. V_1 is the spectral bands of the hydrogen atom here on earth, C_0 is the speed of light on the celestial body, C_1 is the speed of light here on earth. Both formulas have the same starting point comparing the speed of light here on earth, the place of observation, from the speed of light at the point of origin.

The second formula works off the assumption that the speed of light dictates the height of the electron orbits as reflected in the spectral bands or frequency or height of the electron orbits. The atom and its electron orbits expand and contract according to changes in the speed of light. Both formula's deal with the differences in the speed of light but have different ending points. If you apply both astronomical observations to the same source of starlight in the cosmos and get the same results for the starting point, the two observations confirm, validate and reinforce each other and you can be sure of the measured differences in the speed of light.

Observing differences in red shift according to the differences in the speed of light would explain the observation of red-shifted light from the centers of nearby galaxies. Remember the red shift indicates a lower speed of light at the source for the light. It stands to reason that the gravitational potential of a galaxy is higher at the center of a galaxy and it also stands to reason that the higher the gravitational potential the lower the speed of light. It is not possible to observe light from the center of our own galaxy, the Milky Way. But observations of star light from areas close to the center of the Milky Way also show a red shift of star light from that region. Hence the red shift of star light appears to coincide with increased gravitational potential at the centers of nearby galaxies and star light from near the center of our Milky Way. It further throws into question the extreme red shift of Quasars and the idea that any red shift in of itself is due to motion. If the Universe is an aggregate, not one coherent entity, then the gravitational potential and speed of light is different in different areas of the universe and it produces different wave lengths in observed light.

If this is true then you should be able to locate violet shifted stars and star light from the outer edges of our own Milky Way. Violet shifted light is indicative of higher speed of light at the source. The position of these stars would be relatively static to us. As the Milky Way rotates as one unit and you can't argue that the star on the outer spiral arms of the Milky Way are closing with us at a great enough rate to produce the calculated violet shift. Find violet-shifted star-light from the outer edges of the Milky Way and you have proof that the frequency and wave length differences in star light are due to differences in gravitational potential and the speed of light not motion.

Another argument for the big bang theory is the so called, red shift of galaxies increasing as their distance from the Milky Way increases. The argument is as follows: the greater the distance from the Milky Way the greater the red shift. This implies that due to the conventional wisdom where the red shift is the result of motion, the further you get from the Milky Way the faster the galaxies are moving away from us. This is considered to be proof of the big bang, because the motion proves the universe is expanding. It may simply be that the speed of light is higher here on earth, than the average speed of light in the universe. If you equated a red shift to motion, you distort your observations. The same observations made with different assumptions produce different results.

Within the past two decades they have discovered the so-called motion-induced red shift is accelerating. By equating the red shift to motion, they are arguing that the expansion of the universe is picking up speed. Nobody has ever explained what force is causing this accelerated motion. Newton's laws of force and motion have never been demonstrated to be wrong. What force is it that is pushing the galaxies and the universe apart at an increasing rate of speed? It is an interesting question, Yet, the observed motion as supposedly proven by red shift of star light from stars and galaxies outside the Milky Way is increasing at an accelerating rate.

Remember how a change in wave length due to differences in the speed of light in our galaxy would cause an apparent red shift in the star light from distant stars, leading to an apparent red shift. There is an inverse relationship between G and the speed of light. As G goes down the speed of light goes up. If the universe is not in a runaway expansion, that is, if it is not being pushed outward, at an ever-increasing rate. It is possible that the so-called acceleration of the red shift can be explained as a drop in G in the area of the Milky Way as space is being stretched by tidal forces of our fall to the great attractor.

<u>What is the great attractor?</u> In the seventies and eighties of the last century it was discovered that all the galaxies and stellar formations for billions and billions of light years around the Milky Way were moving in the same direction. The question arose, moving towards what? This became known as the search for the great attractor. Scientists put up a microwave observatory by high attitude balloon and determined the direction of the collapse with great accuracy. Once they had the direction, they began to search for the cosmic formation that was attracting so many galaxies. It was Harvard University that determined the collapse was in the direction of a "star wall" a massive formation of stars and galaxies spanning hundreds of millions of light years.

The acceleration of the red shift is because the value for G here on the earth is dropping, increasing the speed of light with a corresponding red shift in the wave lengths from other areas of space with a higher value for G and lower speed of light. As a three dimensional and relatively static entity, all galaxies would tend to collapse or be drawn to the most intense gravitational source in their area of the Universe. This is exactly what is happening to our galaxy as it collapses toward the Great Attractor, the Star Wall, discovered by Harvard University in the 80's of the last century. This fall toward the Star Wall is increasing the red shift due its effects on G and the speed of light here on earth. Gravitational tidal forces are decreasing G in our area of the universe as we fall towards the great attractor.

Drop three weights one second apart from a great height here on earth and the distance between them will constantly expand. The first weight will accelerate away from the second at a constant rate of 32 feet per second and accelerate away from the third weight at a rate of 64 feet per second. The longer they fall before hitting the ground the more the distance between them will be stretched. As we fall to the Star Wall, the distance between our galaxy and the other nearby galaxies is being stretched in a similar fashion. Because gravity is not propagated instantly. The distance to and from the great attractor of the other galaxies in the universe is a function of space-time.

The density of cosmic mass in the great attractor has and is growing over time. As it does the increased rate of attraction reaches the affected galaxies at different times. Like they were dropped at different times. The slower the rate of gravitational propagation the more pronounced the effect will be over time. The ratio of space to mass here on earth is increasingly dropping G and increasing the speed of light and the apparent red shift.

I am not saying there is no such thing as motion-induced doppler shifts. The issue is not that clear. I am aware of pronounced doppler shifts in binary stellar formation orbiting each other at high speed. Indeed, if the great attractor, the star wall, is an area of much higher ratio of mass to space, how come the star wall is not red shifted? The answer is that our fall toward the star wall, our motion toward the great attractor, produces a violet shift in its radiation canceling out the red shift. Our fall to the great attractor and the fact that space is being stretched by that fall contributes to changes in G, producing an acceleration in the red shift and by the very nature of physics the cosmic red shift will continue to accelerate, as the distance between us and our surrounding galaxies continues to grow and G drops in response.

If the doppler shift of star light is due to difference in G due to the fact that different areas of the universe have different space to mass ratios and a different speed of light. Quasars are areas of the universe with much higher mass to space ratios. Areas of the universe composed of extreme densities of mass. Galaxies that have condensed into formations of blackholes orbiting each other with remarkably few stars left or they maybe formations like the great attractor itself. Their motion may not be all that pronounced.

This leads to a view of the universe I refer to as the Swiss cheese universe. An infinite block of Swiss cheese whare the cheese is cosmic formations and dust clouds in areas of space with a relatively low mass to space ratio. The holes are super dense cosmic formations composed of galaxy clusters that have sucked the surrounding area of the universe dry, rendering the surrounding space relativity devoid of mass. Eventually the black holes in these super dense areas explode creating hyper-massive dust clouds billowing outward, filling the void and starting the process of stellar and galactic formation over again. In an unending cycle.

I assert now that the accelerating red shift is due to a drop in G changing the speed of light here on earth due to our fall towards the star wall or great attractor. Quantum Gravity forces one to conclude G is variable. Countless measurements have shown G is different everywhere you measure it, If G were a field force it could well be a universal constant. The exact same everywhere, but it is not. By the very nature of falling into a gravitational frame of reference, the distance between you and objects close that fell first and objects further from the gravitation source grows as you fall. Altering the mass to space ratio. Altering G and the speed of light. These are some of the considerations raised by the accelerating red shift, which is due to a drop in G and the result of the fall of the Milky Way toward the great attractor. Clearly one can form an informed argument that G is dropping due to our cosmic motion towards the great attractor. G is dropping and the speed of light is changing. We can be fairly certain of that.

Let us talk about the practical implications of the drop in G and change in the speed of light. The drop in G would have the effect of expanding the solar system and increasing the speed of light and it has been going on for centuries. The speed of light was originally measured by the astronomer Roemer. The figure is said to be off because the measured size of the solar system was wrong. It is sheer human vanity to assume we know better now, because Roemer was a hack. Roemer was good and his measurements were quite accurate. Repeat his measurements and you will find the speed of light has expanded since it was declared to be a fixed and constant 299,792 kilometers per second, in the twenties or thirties of the last century. Especially when one accounts for the expanded size of the solar system.

Measure the size of the solar system and you will find it has expanded since the figures were agreed upon fifty or sixty years ago. The size of the solar system would expand as G drops. The solar system was literally smaller when Roemer measured the speed of light. The periodic eclipse that Roemer used to measure the speed of light has changed as well. The moon is receding from the earth. The laser measurements of the distance to the moon prove that, but these figures are off, due to the increase in the speed of light.

Remember the speed of light is variable, but according the dictates of relativity it will always measure out at 299,792 kilometers per second in the small domain. Despite the fact it can be measured differently in the large domain. Physicist and astronomers have to wrap their head around the fact the speed is variable and changing, while at the same time according to the dictates of relativity it will always measure the same here on earth.

Einstein was misunderstood. The speed of light is not constant. *The measurement of the speed of light in the small domain is constant.* Measure it in large domain and you will find it is different with in the solar system. If astronomers measure the speed of light the same way Roemer did, while accounting for the expansion of the solar system, they will find it has increased over its previously recorded earth-bound value. Relativity is scientific law and everybody has to understand what that really means in practical terms. That realization is that the speed of light is variable, while its measurement here on earth will always be constant. People have to grasp that essential contradiction. Relativity is scientific law. This means while the speed of light can only be measured at 299,792 kilometers per second here on earth or any clestial body, it is in fact vairable and actively changing and can only be measured differently in the large domain.

All in all, one is forced to consider the idea that the Universe is both infinite and eternal, with no beginning and no end, in size or time. There was no specific act of creation. The universe has always been here and always will be. There is no expansion of the universe from the big bang. It is three dimensional, with galaxies moving towards other areas with a greater ratio of mass to space. While other areas are expanding from the explosion of massive black holes. Producing massive dust clouds and leading to the formation of new galaxies in motion. There was not one big bang but several smaller explosions. An infinite number of explosions in the eternal life and existence of the universe. A universe without an end or beginning, in size or time.

Is it easier to believe the universe started out as one mass in a void, which exploded for unknown reasons and the debris is now in a runaway expansion picking up speed as it goes, despite the fact that our part of the universe is plainly collapsing to an extremely massive stellar formation or is it easier to believe the universe is not one coherent entity with a constant speed of light but rather an aggregate with variable speeds of light in differing area of the universe with different values for G.

Again, the Universe is both infinite and eternal and literally has no beginning and no end, in size or time. Portions of it die as they are absorbed by massive black holes and are reborn as these black holes explode outward, forming massive dust clouds which condense into new galaxies and start the process all over again. This conclusion fits all current astronomical observations and is compelling if one examines the basic issues of cosmology. To assert the Universe was born out of one unbelievably dense mass raises the question where did that mass come from? If the mass was always here what changed to make it explode? Logic compels one to that say something must be eternal. Please consider the distinct possibility that, the universe is infinite and ageless. That the universe is both infinite and eternal with no beginning and no end, in size or time.

Humanity and the Divine

Belief

"Knock and the door will be opened"

a saying of Jesus Christ

I grew up an atheist, thinking that people who believed in God were fooling themselves. My belief in God and human spirituality began at age 22 when I had a dream that came true. That was impossible and really had a major impact on me. All total in my 64 years I have had less than a dozen dreams that came true and it is extremely rare for me to dream the future. My dreams are normally fantasy. When the dream came true at age 22 it was an epiphany, a defining moment in my life. I became a Christian at age 53, when a woman I was living with was healed of a ruptured spleen by prayer. This convinced me that Jesus while he may not have walked on water, he truly did heal. The belief Jesus was a man born of natural birth and did heal is known as Gnosticism. Have no doubt about it this author is a Gnostic Christian.

Ask a psychiatrist, they will tell you that belief in God is not irrational. Being rational consists of dealing with the daily struggles in life without obsession or compulsion. If you hold down a job, have a hobby, date someone significant in your life or are married with or without children, it does not matter if you think God did or did not part the Red Sea. Sanity and rationality consist of dealing with life, without undue self-inflicted handicaps.

A psychiatrist will also tell you that belief in the supernatural power of God is not in or of itself irrational. You are not a deluded fool if you think Jesus was born of immaculate conception. What exactly you believe about Jesus, Mohammed, Vishnu or Moses is a personal choice irrelevant to your sanity. Your belief or non-belief in God is very much a part of your self-image or superego as psychiatrists call it, but it is the way you deal with the emotional content of your life that defines how rational you are. You can be a Buddhist and not believe in God at all and be a good and decent person.

I am a Unitarian. We don't believe anything, we discuss it. We don't practice religion, we study it. We study morality and religion. In over fifty years as a Unitarian, I have never once met a Unitarian who was concerned with salvation. Yet in a Unitarian church you will find healthy debate about what is good and what is evil. We are interested in the struggle between right and wrong in life. Both in society as a whole and as individuals in our role in society. The Unitarian religion is sometimes dismissed as organized atheism. I know of Unitarian ministers who are out and out atheists. But in the church where I grew up and attended services, after services there were two tables in the lobby. One had a sign on it, "The Atheist Corner" the other had a sign, "The Christian Corner". Religious belief is a very personal thing.

EVOLUTION

"The master's crafting hand"
a phrase referring to God used repeatedly by Charles
Darwin in his book "The Origin of the Species".

Before one criticizes the theory of evolution, they should read Charles Darwin's *The Origins of the Species*. It is a towering intellectual achievement that explains evolution as the natural order of things in our world. It describes the endless competition and struggle of life forms to prosper and reproduce and how it produces change in the variety of life forms that surround us in the natural world. Charles Darwin was astute and keen observer of natural world and it is the body of his observations that make *The Origin of the Species* such a convincing masterpiece. Charles Darwin got the idea for the endless competition and struggle of life that causes evolution from a book written by a Roman Catholic clergyman about the endless struggles of humanity. While Charles Darwin believed in God his work produced an argument in our human society about the relationship between humanity and God, that is still going on today.

Life is an imperative of the universe. Radio astronomers have discovered amino acids in deep space. Comets contain viruses. Deep in the oceans near volcanic vents in the superheated water scientists have discovered chemical life sacks of chemicals and minerals anchored to the seabed, which are shaped like giant worms. These organisms grow chemically without benefit of DNA. Having more in common with the chemical process of photosynthesis in plants than any other life process. It was probably the fusion of this form of life and free viruses that created the first living cells from which all other life evolved. The variety of life forms here on the earth is astounding, but there are other planets in the universe and I am certain that there is such a thing as intelligent ammonia life on planets where the average temperature is below zero, the freezing point of water.

The human race is just one of an almost infinite variety of life forms here on earth. There are millions of species here on earth. Evolution is part of the natural order of things, a natural order ordained by God. The universe is an infinite and eternal engine of life and death. For 170,000,000 years dinosaurs ruled this earth. Humanity began its descent from an upright walking ape about five million years ago. The entire human race is in the same position as most children, a happy accident and not a planned birth.

Human life has its origins in the animal kingdom and humanity is still struggling to rise above our animal origins. Human nature itself comes from our evolutionary history. Our drive to find some purpose or meaning in life has also evolved as society has evolved. Our instinctive drives to eat, communicate, reproduce and survive are all products of evolution. This drive to find fulfillment is the product of social evolution. It is a conflict between our ability to think and comprehend and our drive to live, to exist and endure. We fear death and we want to have victory over it. We want to leave something behind which death cannot alter, it is the intellectual equivalent of marking our territory.

The human species is a pack animal. We form groups and share the activities and work of survival. The human species has by far and away the most complex social interactions and social structure of any species. Our society is dependent on our language abilities, the ability to communicate with each other, but has its roots in our origins as small bands of hunter-gatherers. For millions of years the human race lived in small bands of about 30 to 50 individuals, surviving in a territory about half the size of Manhattan. This was the setting in which we learned to work together as a team on the hunt and communicate with our young teaching them the skills of survival.

Darwin emphasized the importance of sexual selection in evolution. The average male of our species is significantly larger than most females. This happens in a species where the males fight over reproductive females. Early in human evolutionary history males fought each other for reproductive rights and kept harems. The human male has an instinctive need to prove his dominance. To this day teenage boys are prone to fighting as they enter a period of sexual maturity. As teenage boys grow older, they learn to channel the urge to dominate into group activity and the desire to be the leader or star of the team. Men still fight and society rewards the victors with glory and acclaim. This entire aspect of our society is a holdover from when in our evolutionary past males fought because of the need to win females and reproduce.

Our sexual ethics are the result of evolution. Men are much more sexually promiscuous than females. Men are instinctively driven to mate with as many women as possible and sexual promiscuity is more tolerated in men. A woman's attitude toward her mate is not always shaped by exclusivity. Polygamy still exists in this world and even in societies where polygamy is banned many women don't care what their man does away from home so long as he doesn't bring a disease back. A venereal disease is a threat to her own reproductive capacity. There is an age old saying "There is a lot of difference between other women and another woman." By this logic a woman doesn't care if her man has sex with other women but if he has an affair with one woman and makes an emotional investment in her, that is threat to her relationship with her man and the needs of her children. Men do not reciprocate the attitude. If a woman has an affair outside the relationship, it is seen as a direct threat to the relationship and his parentage of any children as a result of the relationship.

The two sexes have different attitudes toward reproduction. A man seeks to mate with as many women as possible. A woman is instinctively driven to seek out only one father for her children. Her judgment of what constitutes a suitable mate is affected by status and material circumstances as a woman looks for a man who can provide and is a success in society. Surveys of women in the United States consistently find the wife of the president is almost always the most respected woman in society, because she is the mate of the most successful male. Physical beauty is also a consideration for both men and women. This is because symmetry and good features, both of which are major factors in beauty, make someone a better mate for reproductive purposes. The sexual desire for beauty helps weed out the defects in the species.

The fact that women seek only one mate is not social upbringing: it is hard wired into instinctive drives of woman. This hard wiring of the sexual partnership for the woman is illustrated by studies of women's sexual fantasies and behavior. It is a common male fantasy to have sex with more than one woman at a time. The fantasy is quite common and most men will jump at the chance to do this. Two thirds of all women have similar fantasies of sex with two or more men, but are very reluctant to do it. One in six women actually try it. Of those that do fully ninety percent find the experience so emotionally upsetting they never do it again. If sex, was just sex and nothing more for women, not a biological imperative, multiple partner sex would be far more common in women. Fully ninety percent of all women find multiple partner sex, instinctively unacceptable.

This woman's bonding with one man for parentage of her children is manifest in other ways. The woman is intellectually and instinctively driven to be faithful to the father of her children, this is considered a responsibility of the relationship and when a man terminates a relationship where there are offspring, typically the woman sees it as a betrayal of that trust and will engage in promiscuous behavior to punish the deserting mate

Females seek to build a relationship and make the male commit to her and their children. Of all the species on Earth humans make the biggest biological investment in their offspring. Pair bonding better known as "love" is a biological fact. Men and women commit to each other and their children in love in order to share the burden of child-rearing. Generally, in most societies' men are seen as the providers of the resources of survival while mothers are the primary caregivers. Children are almost totally dependent on their parents for survival for the first twelve years. Only in adolescence do they begin to become self-sufficient and breakaway from their parents, a process that takes several years. In the animal world most species' offspring are on their own in less than a year.

Love for your partner and your children is a biological fact. Instinct rules. Love is not a learned response. Nobody needs to write books about how to fall in love. Your bonding with your mate and children is natural. When it comes to a mate the hormones do it for you. Some people do come together for practical reasons and learn to love each other, but even in these relations the hormones and biological responses come to dominate and shape the relationship. Put more simply arranged marriages are still a fact in this world, but a man still needs to get an erection to become a father. It takes a very emotional stunted person to go through life without falling in love.

The needs of the male to solicit sex led to profound changes in our biology. Upright walking became an important part of the male sexual display. Penile size also grew to excite the female. The human male has the largest penis of all the apes. In most species the penis is in a sheath and is only visible when aroused. In all the other apes the penis is in a hair covered sheath. The human penis is hairless and plainly visible when a male is naked. Sexual display played an important part in the development of human's relatively hairless skin. The females partially lost body hair and developed protruding breasts as a form of sexual display. The enlarged breasts of the human female are not biologically required for feeding of children. Female chimpanzees nurse their young quite well despite the fact they have no protruding beasts. The protruding breasts primarily serve the function of sexual display to excite and solicit the male's sexual attention.

Human sexuality has been shaped by the needs of pair bonding and group survival. The urge to put the safety of women and children first in times of danger is instinctive. The males who do this are protecting the future and reproductive potential of the group. In our evolutionary past men fought each other for reproductive rights. Early in human evolution the dominant male would drive out the young males and mate with the females when they were ovulating. As the demands of child rearing grew and pair bonding became necessary and human female biology adapted, the hymen grew to help hide the scent of the approaching fertility giving the female more choice in who to mate with. Studies of married couples found the males were quicker and easier to arouse and were more jealous when the women were ovulating.

If two women share an apartment or house their menstrual periods will change and synchronize so that ovulation occurs at the same time. In the troop of early humans this meant all the women were fertile at the same time making it very difficult for the dominant male to impregnate more than one female at a time. The female human is also unique in that she shows no visual sign of ovulation or fertility and the male is unsure of his parentage of any offspring unless he is the only male sexually interacting with the female. Most species only mate for reproduction when the female is fertile. Humans are an exception to this rule. Repeated sexual intercourse became a requirement of pair bonding.

Young adolescent girls get the urge to mate around twelve or thirteen. This manifests' itself in something commonly referred to as the first crush. They become infatuated with some man or boy other than their father. In modern society with its media driven culture this maybe some male figure on the television or in film. In the 1960s, millions of teenaged girls around the world went hysterical over a rock and roll band known as the Beatles. It is an example of how modern society effects behavior, but evolution and human nature still rule.

The human species stands out from the other animals on the earth in the variety and complexity of our tools. While some other species make tools, the complexity and sophistication of our tools is an achievement which no other species on earth even comes close to approaching. Increasingly human society is defined and shaped by our tools, but it is still our social interactions with others and group activities that give us our sense of fulfillment. The solution for the anxiety and depression so common in society today is not drugs such as Prozac, which is really another tool in itself, but rather social involvement. Don't just go to the gym, join a club or some sort of team. The cure for many psychological ills is to get involved socially in an activity that builds friendships and social relationships. Nobody is an island. Bonding with others in a shared activity is a biological and natural cure for many socially induced emotional troubles. It taps the instincts of our common human nature as it was bred into us by evolution to ensure our survival. Ultimately our sense of worth comes from others, the recognition that we have a part in our society.

We also gain reassurance from the idea that there is a higher power, that as individuals we have some worth, in the unending cycle of life and death here on earth. Our intellect compels humans to look for this sense of worth in philosophy and faith and ultimately it is philosophy and faith not our tools, which separates us from the animals and will lift humanity up above its purely animal origins.

Human spirituality

"I know I have a soul that goes with me when I die"
the philosopher Socrates

Intellect is among the most respected attributes any person may possess, but the ultimate measure of a person is not their ability to reason, but rather their character and the ultimate measure of a society is not its technology achievements or its wealth, but rather its social justice. The American society is the richest society in the world and leads the way in science and technology, yet it suffers from a rash of young males murdering each other in its schools and streets. A major part of this problem is the false intellectual assumption, that science and evolution preclude the possibility of God. There is an existing unspoken intellectual prejudice against the idea of God and this has a crucial cost in society. If there is no ultimate right or wrong, isn't anything permissible? This sort of nihilism is the outgrowth of the clash between the new ideas of science and the old dogmas of religion. Many people trust science more than they believe in the dogmatic concepts of religion and science has failed to fill the moral void left by this intellectual transformation.

Charles Darwin's *Origin of the Species* is a towering intellectual achievement and evolution is without a doubt the natural order of things. It is however sheer human vanity to maintain that God created the earth as the specific abode for the human race or assert that humanity is God's greatest creation. As such the proper intellectual response to "Origin of the Species" is not to dismiss God, but rather to apply the same intellectual diligence that Darwin had to the search for evidence of human spirituality and God in the realm of human experience.

What would such evidence consist of? Are there things in the human experience, which are supernatural in nature and defy science? As science has progressed rational explanations have been found for many things once thought to be mystical or supernatural in nature. Sneezing used to be thought of as the body expelling demons. This is why "Bless you" became the common and acceptable response when somebody sneezes. The very term supernatural has come to be associated with hoaxes and imply superstitious thinking or beliefs, but there are common or ordinary events in life which defy scientific explanation and are supernatural.

People think science is perfect, some people have gone so far as to say mathematics is the only real truth in this world. The theory of metrically elastic curvature of space is mathematically correct and is trash. If you think mathematics is perfect, attempt to accurately calculate the surface area of an egg. An egg is a common ordinary thing, which defies mathematical description. You can literally write a book about the egg and never come close to discovering how to mathematically calculate the surface area of one. Just like the egg defies mathematical description, there are things in life that defy scientific explanation things that are truly supernatural in nature and can be considered evidence for human spirituality or God.

There is evidence of human spirituality in psychic experience. I grew up an atheist. My belief in God began when at age twenty-two I had a dream that came true. The dream I had was of a total stranger I had never seen before telling a friend of mine he was getting a divorce. The person was exactly as I pictured him, the body language and the words the stranger used were all exactly same as in my dream the night before. That dream was impossible.

I read Einstein at some length at a young age and Einstein never said time was a dimension, but rather a coordinate of observation and a relative one at that. For time to be a dimension would require billions of parallel universes all occupying the same space. We live in a solid three-dimensional world and occupy space. Time is not a dimension. My dream defied every known law of science and was clearly supernatural. That dream was the beginning of my belief in God and fate—the beginning of my conviction and belief that there is more to life than just living and dying.

Dreams are how the mind processes need, desire or conflict. Successful people sometimes dream their successes, a key presentation or how to solve a problem confronting them. Athletes sometimes have dreams of new moves or actions they can take in a sports contest. Juicy Lucy my teenage phoozball partner used to dream up new phoozball shots. Sometimes however dreams of the future take on a predictive quality about events beyond our control, which is rarer and more difficult to explain.

Like Sitting Bull's dream of soldiers falling upside down into the American Indian encampment before Custer's defeat at the Little Bighorn. The Indians inflicted a stinging defeat on one of the three armies hunting them, after Sitting Bull's dream though Sitting Bull was certain this victory was not what his dream was about. Sitting Bull's dream only came true when Custer in an act of gross military stupidity divided his forces into three separate groups of slightly more than 200 hundred apiece and attacked Sitting Bull's village which contained nearly ten thousand warriors.

Perhaps we are all given to dream the future at times in our lives. Psychologists who have studied the phenomenon of Déjà vu, attribute it to people who have dreamed an event before they experience it. How common dreaming the future is in general population is something that is best studied by trained psychologists. If you study people's dreams one should separate or differentiate between the dreams that address need or desire and those that are truly predictive. The predictive dreams are extremely rare. After all we dream every time we sleep. Most dreams are in response to need or desire. But people sometimes have predictive dreams, which are outside the realm of need or desire. Are we all given a glimpse into our own spirituality as a part of our gift of free will? By this means will we know there is something at stake in the choices we make in life? Talk to other people about the dreams you remember and see what you think. It is an interesting fact that the culture with the lowest rate of mental illness in the world is a Polynesian society whose members discuss their dreams with their family every morning at breakfast.

There are true psychics who know some little specific details about other people's past or future events. The Catholic Church studies psychic predictions and records them. A lot of people dismiss psychics and psychic phenomenon, because of how psychics are portrayed on television and in film. There are also a lot of fake psychics who are really con artists. To separate the true and fake psychics ask what is their gain or stake in what they're saying. Never trust a psychic who takes money for their efforts.

I know a true psychic, who wishes to rename nameless. Her gift is believable because she does not take money for it or profit from it in any fashion. She also does not work for the police, because she believes that if you push it and abuse the gift you destroy it. She uses it as a conversation starter.

One incident comes to mind. She was sitting by the door in a bar having drinks with her family and friends when a man walked into the bar and she said to him "There is something wrong with your knee and you just went back to your wife." The man confirmed he had just had knee surgery and had gone back to his wife after a separation of six months. Where is the deceit or subterfuge in talking to a total stranger in a bar?

There is a famous case of a psychic who worked for the police in Chicago telling a detective investigating a case in which a person had gone missing from a commuter train, "The person you are looking for is dead and in water.", said the psychic. It turns out the man had gotten drunk with his friends and in a drunken state attempted to get off the train while it was moving, fell into a river and drowned. How do you fake things like that?

One very famous psychic was Edgar Casey of Virginia Beach, Virginia. He had a very rare gift. He had no medical training, but could just by laying hands on a person's medical file diagnose their illness without actually reading the file. Sounds impossible I know, but he did this accurately and effectively over a hundred times.

Jesus was psychic at the last supper after Judas left to betray him, Peter said "I will never betray you, Lord." and Jesus replied "Before the cock crows you will deny me three times." and it came to pass. This is the essence of the psychic gift—an ability to have specific knowledge of some small fact that doesn't reveal the ultimate picture and nature of God's purpose. Jesus did not understand God's ultimate purpose he only knew he would suffer and that Peter would deny him. It is an example of limited specific imperfect knowledge, but knowledge that is none the less impossible to possess without some form of human spirituality. This psychic ability is given to some more than others but is perhaps given to all of us to some small degree in our Déjà vu dreams of the future.

There is evidence of human spirituality beyond the grave in After Death Communications. A wide variety and range of these After Death Communications (ADCs) are explored in the book *Hello from Heaven*. This book is a very diligent effort and makes for interesting reading. The earliest record of an ADC comes from a Roman essay "On Divination" by Marcus Tullius Cicero who died 40 years before the birth of Christ. He recounts how two friends from Arcadia traveling together separated after they reached Megara. One of them went to the inn and the other accepted the hospitality of a friend. The friend and his guest finished the evening meal and retired. In his slumber the guest was visited in a dream by his traveling companion, who told him "The innkeeper has murdered me, flung my body in a cart and covered it with dung. Please I beg of you be at the gate early in the morning before the cart can leave town." Stirred by this dream, the man waited at the gate and confronted the man driving the cart, who fled in dismay. The man recovered the body and reported the murder to authorities and the innkeeper was duly punished.

While sixty percent of people, experience some form of communications and reassurance from dead loved ones or significant others, most commonly in the form of dreams, there has never been a scientifically proven medium, someone who can talk to the dead. Harry Houdini the famous escape artist devised a code with his wife by which he would attempt to communicate through a medium. Despite visits to dozens of different mediums, the experiment failed.

OF SCIENCE AND GOD

Some people in medicine and science refer to ADC's as After Death Hallucinations. But some of these communications are compelling and very hard to dismiss particularly those where the person does not know the other person is dead or they communicate some fact that could not be otherwise known, as was the case with the man and his murdered traveling companion.

There was a sect of early Christians known as the Gnostics. The Gnostic Christians maintained the resurrection was of the spirit and not the flesh. Who asserted the spiritual communication of Christ to his disciples on the Pentecost, imploring them to carry on. Jesus' after death communications also include his appearance before the apostle Paul on the road to Damascus.

All in all, you cannot simply dismiss all ADC's as hallucinations it just doesn't wash and they directly reflect on human spirituality. After death communications are evidence of a human spirituality that survives death.

Reincarnation and the immortality of the soul - Tibetan Buddhism and Hinduism are two religions that assert the soul is immortal, born again and again. Immorality of the soul raises an interesting question. Hell might be eternal torment, the ultimate punishment, but what are you going to do in heaven for billions and billions of years? Don't you think you might get a little bored? Christian preachers commonly say we will all sing God's praise in heaven. You could sing every hymn of praise for God ever written, write a few thousand of your own and still not come close to using up the first few million years, much less a billion and a billion years does not even begin to touch eternity.

People who believe in reincarnation often point to the recounting of memories of past lives recovered under hypnosis. This argument doesn't hold much water. It is a common practice for the hypnotist to tell the person to be hypnotized, that they will remember past lives and recount them. A hypnotic subject has a great desire to please the hypnotist and it can easily be argued that the person is engaging in fantasy to meet the preconditioned expectations of the experience and please the hypnotist.

There is however a famous case of an eight-year-old boy in India traveling alone to a town he had never been in. He claimed to be the reincarnation of a specific man, calling his supposed widow by her correct name and identifying other people whom he'd never met by their correct names. The boy's family did not profit in any fashion from the incident and it is the kind of incident that would be difficult to imagine any means or motive for the boy to fake it.

What little is known about life after death, comes from people who have experienced clinical death and have been revived medically. Their stories are quite consistent, moving through darkness, into a bright white light and communing with dead relatives and loved ones. There are books that document the common experiences of people who have died and been brought back to life. One is titled *Life at Death*. The phenomenon of people who have clinically died communing with dead loved ones has been testified to by many people of no particular religious beliefs. Most of the people who have testified to this phenomenon are not especially devout in any religion. While these studies have primarily taken place in western democracies which are mainly Christian in makeup. The people who have testified to this phenomenon do not go to church every Sunday and this experience while it does make them believe more strongly in human spirituality. It almost never leads them effusive praise of the lord. Apparently, it appears God does not require people to worship him to extend to them his mercy. Evidently God in his infinite mercy allows <u>all good people</u> to commune with loved ones in death, ensuring their love endures beyond the grave.

The intellectual argument against this is that as the brain begins to lose oxygen the person hallucinates. The counterpoint to this argument is that, since every drug induced hallucination is different, why are the hallucinations induced by death so consistent, unless they comprise a true and valid experience, a glimpse into the ultimate fate of the human soul?

These are some general observations about some common human experiences, which support arguments for human spirituality. Nothing about human spirituality can be proven in scientific terms. All we can do is point to those experiences' science cannot explain and wonder.

GOD

"I am that I am"

God to Moses from the burning Bush

There is scientific evidence for God in the sudden and dramatic remissions of illnesses, which are scientifically verifiable, but scientifically and medically unexplainable in nature. They are called miracles. It is the sort of thing where a believer needs no explanation and for a non-believer no amount of explanation is enough. Some people will deny that any miracle is the work of God because they cannot explain how God did it, refusing to believe what they cannot understand. A miracle like God is not self-evident, a sign does not appear on the forehead of a healed person saying "Healed by God". The persons look normal, so one is not inclined to believe that anything extraordinary happened to them.

How do you apply science to this debate? A medical condition as established by test results, X-rays, MRI's, CAT scan, or others form of technological or even visual observation is an empirical fact proven by evidence. The sudden remission of such a condition, when a remission occurs, is also an empirical fact proven by evidence. The sudden and dramatic remission of an illness or condition that is an established fact,

Of Science and God

is empirical evidence of something supernatural. It is only an intellectual prejudice that makes us doubt God.

Things do not just happen: when something happens something makes it happen. We live in a world of cause and effect. If something changes, something made that change happen. Despite what people may think, even random acts have a cause. Something made that random event occur. Medical miracles are proof of the supernatural in human experience, it is impossible to consider them and not reach such a conclusion unless one is engaging in denial.

Do not confuse legitimately documented medical miracles, with faith healers. There are a number of Christian ministers and laymen who claim the power to heal. Not one of them has ever produced a medically verified miracle. Typically, these people work tent revivals, but even some major and respected Christian ministers like Pat Robertson founder of Christian Broadcasting Network have at times engaged in this exercise. Bless his heart for trying.

The science did not exist to verify the miracles of Christ and there has never been a legitimate scientifically verified hands-on healer. It is, however interesting to note a recent scientific study of illness and prayer. It found that people who have family members and loved ones praying for them have a higher recovery rate and live longer. There are however medically documented remissions of serious illnesses, which defy all rational attempts at explanation and are truly supernatural in nature and qualify as miracles.

One miracle that made the news over a decade ago, was the disappearance of a stomach tumor from a young girl in India after the family prayed to Mother Theresa for her intercession. We live in a solid three dimensional world of cause and effect. A tumor is a real thing. How can a stomach tumor be there one day and gone the next? How can a tumor be there on one X-ray and gone the next? This tumor spontaneously dissolved and was replaced by healthy tissue. A tumor has a different chemical makeup than the healthy tissue around it. A tumor has mass and can weigh a pound or more. A couple of pounds of Uranium dissolved into energy over Hiroshima and over 200,000 people died. Yet when this tumor dissolved, the electricity didn't go off. The magnetic didn't spin. The girl was not lifted off the bed, no heat or light was generated. The surrounding tissue was not disturbed in any fashion and healthy new tissue was woven into the intricate fabric of a human body. It is the delicate crafting hand of an omnipotent God.

If you do not trust the miracle I have cited because it is claimed by the church, I will give you a miracle not claimed by any religion. In 1992 shortly before the presidential election. NBC nightly news ran a story of a man who fourteen years before was on his death bed from advanced leukemia and got up and walked away from his deathbed without a trace of the disease. In leukemia, as in all cancers, the DNA of healthy cells mutates and the cancer begins to duplicate out of control in a malignancy. Cancer cells have a different chemistry, for somebody to be suddenly cured of leukemia and restored to normal health the body must somehow be purged of its malignant bone marrow, with healthy cells replacing diseased ones at a fantastic rate. The man did not become anemic or septic from the death of the white blood cells and cancerous bone marrow. The man's blood chemistry went from terminal to normal, suddenly without an explanation. It is unnatural, to say the least.

The very DNA and chemistry of the cancer cells had to be transformed. This is something we can do with today's technology. Gene splicing and its resultant cellular transformations are present day realities. While we can splice new DNA into a cell, we have not yet learned how to remove defective genes. Gene splicing is done one cell at a time.

Of Science and God

These single cells are either bacteria or the fertilized egg of some animal in a test tube. Ask a geneticist or doctor about the practical problems of doing this inside a living body of a fully-grown human being to millions and millions of cells at once. So, a change is observed overnight. Stem cell therapy takes weeks or months to take effect. If you are talking about the sudden and dramatic genetic change in diseased flesh and bone, you are talking about a miracle.

For fourteen years this man's doctor pursued the man who was cured of leukemia, submitting him to every medical test he could think of, looking for a medically explainable reason why this man's cancer so suddenly and dramatically went into remission. He believed that if he could isolate the cause, he could duplicate the conditions in others and save lives. The doctor evidently doesn't believe in miracles and the intellectual prejudice against God is such that, not one word about God was mentioned in the news report. There is no reason except the unspoken intellectual prejudice against God in society today that they could not have ended that report as follows, "To some a mystery, to others a miracle and proof of God." It would have been a better and stronger work of journalism if they had.

In intellectual circles, science is held to be man's greatest achievement and its noblest pursuit. People worship science without accepting it as a human endeavor, subject to the fallacy of human nature. The only thing in science or science fiction that comes close to approaching the forces in play in a medically provable miracle—the transformation of the elements in a living body—is teleportation. This involves the fictional ability to electro-magnetically map and disassemble a material body, recreating it elsewhere. Science will never achieve teleportation. Teleportation involves the creation and destruction of mass in the laboratory. The average person totally fails to comprehend the magnitude of the forces involved and the extreme precision required to even begin

GOD

to approach this proposition.

You can create the simplest of all elements, hydrogen, artificially. Atomic physicist Enrico Fermi did this in the first atomic reactor, with a Fermi bottle. It was an experiment to measure the half-life of a neutron. The hydrogen atom is composed of a single proton and electron pair in loose relationship. The neutron is a proton and electron tightly bound together. As neutrons breakdown in close proximity to each other they should form hydrogen atoms. Fermi put an evacuated glass bottle in the first atomic reactor and weighed the amount of hydrogen produced knowing how many neutrons would travel through that volume of space in a given period of time. His statistical analysis suggested a neutron half-life of two weeks. Better observations have given a neutron half-life of 90 seconds, meaning less than one in 100,000 interactions of neutrons breaking down produced a hydrogen atom and this is the simplest of all atoms.

Creation of a single gram of carbon in the laboratory would require the extremely precise manipulation of billions of times the particles, with sufficient energy to power a small city for a day. Any loss of control in the application of this energy means the machine, directing and channeling it, would go off like a small atomic bomb and the precision required is flat out impossible by today's science. To build a complex atom you would have to manipulate and direct particles to construct the nucleus of an atom. This means you would have to manipulate and direct massive amounts of energy in a space of less than one hundred millionth of an inch. The principle of uncertainty maintains that you cannot measure both position and motion of a particle in that range without effecting and distorting the very things you are measuring.

The teleportation of a living organism would also involve recreating the electrical bonds of the chemicals involved. That involves the

Of Science and God

extremely precise manipulation of the orbit and spin of the electrons within the atom in relation to other atoms in extremely close proximity. Thus, projecting massive amounts of energy through existing atoms without disturbing or distorting them in any fashion. Ask a chemist about the complexity of the molecules that make up living tissue: in attempting to recreate them you compound the problem a million-fold! So, to build a device to create one simple atom or complex chemical molecule one would have to overcome the terminal limits of modern science. We can't weave an atom or spontaneously change the molecular make up of living flesh, but in dissolving a tumor or changing the chemical makeup of diseased tissue and weaving the healthy new tissue into its place, without disturbing the surrounding flesh, that is exactly what God does.

Even if you say simply that there are remissions of serious diseases without necessarily involving God, explaining these remissions defies all known medical techniques, science and logic—it is clearly supernatural. It is something that defies all scientific explanation and rational explanation. It is something which has no place or niche in the natural world cannot be empirically explained or reproduced and can only be considered to be supernatural or something exceeding the forces at play in the technological world. Again, it is only an intellectual prejudice that makes us doubt the supernatural hand of God in these miracles.

There was another miracle reported on the NBC television network. On good Friday, Easter 2008 the NBC Today show aired a news report on a man who was brain dead and recovered consciousness after his grandmother prayed for him. This man had been legally declared dead. His death certificate was signed and his IV removed, when he suddenly recovered consciousness. While he was conscious and aware his recovery was not complete. He still needed rehabilitative treatment. At the time of the report, he was walking with some difficulty. Again, I repeat the man was scientifically brain dead and recovered consciousness. In order for that to happen the dead, atrophied brain cells had to be brought back to life. I know of no other case where someone who was brain dead, has ever recovered consciousness. When I was researching for this book and asked a doctor what exactly would be involved in such a miracle, the doctor insisted that the report was wrong, that what NBC reported was impossible, but isn't that the very essence of a miracle.

CBS news once reported a miracle involving a sixteen years old girl in Australia. This girl had a kidney transplant and was struggling with rejection when the RH factor in her blood spontaneously changed, solving her difficulties. That not only would involve changes to millions and millions of blood cells, but also changes to the DNA in all her blood producing bone marrow cells. There is no rational scientific explanation for how this happened except as a miracle.

Miracles are not always the result of prayer. One under—reported miracle—received wide press coverage but was not described as a miracle. After the reactor in Chernobyl blew up and caught fire. 32 firemen rushed to the roof of the reactor building and poured water on the burning reactor in an effort to contain the disaster. Their efforts were heroic but misdirected. They had no chance of putting out the reactor fire. The smoke and ash of the burning reactor was toxic with radioactivity. All thirty-two firemen collapsed shortly after being pulled off the roof and were taken to the hospital. The level of radiation they were exposed to was so bad, within days all thirty-two firemen contracted leukemia. You can live for years with leukemia. Their leukemia was so toxic thirty-one of the thirty-two firemen died within thirty days. One of those firemen recovered on his deathbed and walked out of the hospital without a trace of the disease.

Understand something about this man's leukemia and the level of radioactivity to which he was exposed. The man was on the roof of the burning reactor. He breathed in the smoke of the burning reactor which was impregnated with particles of plutonium. There is no doubt whatsoever in my mind his lungs had trace levels of plutonium. The radioactivity of this plutonium was what destroyed the bone marrow of his ribs giving him the leukemia. Due to the level of radioactivity he and the other fireman were exposed to, the doctors and nurses that treated him and the other fireman wore gloves and surgical masks to limit their exposure to radioactivity. By all rights this man should have died, but God healed him.

This was in 1986. Obliviously in Communist Russia the press did not attribute the healing of a deadly disease to an act of God. But God knows no borders. God is also aware of the suffering and sacrifice of the men and women who risk life and limb for the protection of their communities. People ask where the love of God goes when the bullets fly or a building catches fire. God is aware and the bravery of those firemen was such that God rewarded one of them with life. God cured him and let him live. I do not know how old the fireman was. He could have died of natural causes since then. But there have been no reports in the press of him dying or contracting cancer since that time. Given the amount of radiation he was exposed to, the fact he lived is a miracle and worth noting as such.

This book was inspired by a medical miracle. As recounted earlier I, the author of this book, grew up an atheist and was an out-and-out atheist until age 22, when I had a dream that came true. This was the beginning of my belief in human spirituality. For the next thirty years my beliefs were very stark. I believed in a God beyond all human concepts of good and evil. As I was fond of saying, people have no more understanding of God than a child does of its parent when the parent is spanking them.

Then at age 53 I was living with an alcoholic woman by the name of Kathy Toler. In March of 2004 she began collapsing unexplainably and complaining of hallucinations, so I took her to the hospital. In the hospital, she went into convulsions and was given an emergency MRI. The MRI revealed she had almost entirely bled to death from a ruptured spleen. She was given an emergency transfusion of four pints of blood, put in the ICU and prepped for surgery in the morning. In the morning, the surgeon came in and asked for an ultrasound of the spleen. The ultrasound revealed a healthy spleen and the surgery was scrubbed. It had to be a miracle. If the MRI was wrong, why did she need the transfusion? If the ultrasound was wrong Kathy would have continued to bleed internally and in a couple of days would have needed a new transfusion. Kathy was released from the Intensive Care Unit and released from the hospital a few days later without further treatment after the doctors were sure the ultrasound was right.

Aware of Kathy's circumstances and the nature of what had happened. This miracle hit me like a bucket of cold water. Over a period of time, I went from believing Jesus was the most successful cult leader of all time to accepting Jesus as a true prophet of God. A man who had a divine mandate. A man who truly performed miracles. I went from dismissing Christianity as an emotional response to the human desire to believe to being a stone-cold gnostic Christian. It took time. It was not the emotional surge of accepting Jesus as your lord and savior. Not the rush and emotional high of salvation. It was a profound and deep-seated intellectual transformation of my beliefs. This is how I came to Christianity at age 53. This book started out as my effort to organize my thinking about religion and its role in the world. Over time the book has become an attempt to convince others there is a God and Jesus was first and foremost of his prophets, while respecting those who believe differently than I do.

OF SCIENCE AND GOD

As medical knowledge has increased the awareness of medical miracles has grown. Two medical miracles are required to beatify then consecrate a saint in the Catholic Church. In the past thirty years the Church has consecrated more saints than it did in the previous 1500 years. While many of these saints where martyrs and as such did not require the medical miracles to be declared saints. In the past forty years the Church has certified well over a hundred medical miracles.

Something that diminishes the church's argument that medical miracles are proof of God, is the church's practice of referring to anything that inspires faith as a miracle. This is played up in the media. How many of us have heard of the woman that baked a cookie that bore the image of Christ and sold it on Ebay? Saying this is a miracle makes a mockery of those medically certifiable miracles, which are truly supernatural. Medicine is the most advanced and highly developed of all the sciences and what it considers to be supernatural can safely be thought of as in the realm of God. Let me reiterate, *medically certifiable miracles, which defy all attempts at a rational explanation, can be considered empirical proof of the supernatural power of God.* I find no conflict between that statement and any principle of science.

Take this book to any college professor who teaches science or a medical doctor and ask them if they can provide a rational scientific explanation of how any of the miracles cited in this chapter happened. You will find that science is not perfect and does not have all the answers, that science has limitations and does not disprove God.

In part, this book was inspired by something that happened in the Columbine High School library where the three teenaged assassins killed most of their victim over two decades ago. One of the assassins asked a young girl "Do you believe in God?" She answered "Yes".

He then asked her "Why?".

When she couldn't answer him, he shot her and then shot himself. I always thought that little girl deserved an easy answer—that supplying that answer might help save other young girls from a similar fate. This book is an attempt to supply that answer. If you believe in God and someone asks you "Why?" all you have to say is "I believe in miracles." One can elaborate further "I believe that there are things in the realm of human experience that science cannot explain. Things that can only be explained as spiritual or supernatural" It is a reasonable rational answer and as such can be a basis for an informed and rational faith in our modern technological world.

What exactly you believe is a personal choice, but there are many well documented medical miracles which are truly supernatural. If God can transform the elements inside a human body, then God can raise a person from the dead or impregnate a woman. God can perform supernatural things. God can defy science and common human experience. The belief in the supernatural power of God is not in or of itself irrational or delusional. It is a question of personal faith as too what you believe in. While I personally do not believe in the Immaculate Conception or bodily physical resurrection of Christ, I do not challenge the beliefs of those who do. Their faith is just as valid as anyone else's.

Once you accept the supernatural power of God many, many things become reasonable. If God can heal, then God can kill, as God did in killing Egypt's firstborn. They used mathematical gaming theory to analyze God of the Old Testament and found, God is a ruthless power player and very jealous of his name. As such it is very believable God let Pharaoh, thought by a nation to be God, pronounce the judgement on his people from his own mouth. One cannot claim science and evolution disproves God. It is simply not so. Ultimately belief is a matter of faith, but don't think evolution or science in or of themselves disprove God, it is simply not that easy or clear cut.

Religion

The Creation
of Religion

"God is Man's Greatest Creation"
a remark by the writer Voltaire

Do not make the mistake of confusing religion and God. God ultimately may be a fact, but religion is a set of practices or recommended behavior founded on someone's opinion about what God wants or what life requires. Like philosophy it helps shape people's opinion about the meaning of life. Religion is a product of human civilization, the result of the removal of humanity from its natural state of grace. Where people live in a semi-developed tribal cultures, myth and legend develop, but people in primitive cultures such as the Aborigines or African Bushman who still live as hunter gatherers. As humanity originally evolved, don't have - highly developed ideas about God. Instead, they live in harmony with the rhythms of nature and never seem to worry about the greater meaning or purpose in life. Two questions which sorely vex civilized societies.

Religion and philosophy are linked. Philosophy analyzes the questions of life and religion seeks to supply the answers. Both grow out of the angst so common to civilized society. The last major philosopher was the Frenchman Sartre. He maintained that all of life is about the search for meaning. This poses many questions for everyone and supplies remarkably few answers.

The politics of religion are often fascist in nature. All religions believe the truth is manifest in their doctrine. The leaders in almost all religions are convinced they speak the only truth that matters. They often think that an idea which is contrary to this truth or opposes this truth is to be suppressed. Politically there is no difference between the Spanish inquisition and Stalin purges. This fact is why religion and government don't mix. Iran the only existing theocracy in the world is doomed to destroy itself with an inquisition or modern version of the Salem Witch trials, the political upheaval that destroyed religious government in the New English colonies of the early United States. So, let us take a look at the roots of religion and its dogma.

CHRIST AND CHRISTIANITY

"Seek and you shall find"
a saying of Jesus Christ.

Jesus was the only prophet of God not to die a natural death and to heal. God's recognition of humanity is implicit in any healing miracle. Christ is Greek for anointed. Jesus was anointed by God to heal as a means to certify his ministry and doctrine. That doctrine can be condensed into a few principles. Do unto others as you would have them do unto you. Forgive others their transgressions against you, so God will forgive you your transgressions. Judge not, lest ye be judged and turn the other cheek. In his middle thirties Jesus came to his calling from God and preached this doctrine for less than three years before he was crucified like a common criminal. His ministry and death laid the foundations for a new relationship between humanity and God. Christianity has evolved over the centuries and it is worth taking the time to look at its roots and the early years of its development.

The story of Jesus birth in Bethlehem and flight to Egypt is probably correct because according to Jewish records Jesus was seen and known in Egypt before he began his ministry in Galilee. Magi was the old term for astrologers who read signs and portents in the planets and the stars. Most religious scholars think the star of Bethlehem was a conjunction of the planets. Possibly a rare conjunction of the planets Saturn and Jupiter where they merge into a single point of light, that occurs only once in every, four hundred years. Why this conjunction of the planets? It coincides with a documented census in the region and the census was how the Romans set the amount in taxes levied on the local rulers and citizens.

How did the Magi find Jesus? Bethlehem was a small town and there was in all probability only one inn and Jesus was born in its stable manger. How did the Magi know Jesus was the new king they sought? The virgin Mary was in all probability psychic and told the Magi things about them or their quest which she could not have otherwise known. What makes me believe this? Jesus was a male psychic and male psychics are almost always born to a psychic mother or have a maternal grandmother or great grandmother who was psychic.

There was a second Jewish temple in Alexandria Egypt where Jesus could have been educated. It was destroyed by the Romans as a result of the Jewish uprising in Israel some forty years after the death of Christ which also led to the destruction of the great temple in Jerusalem. In Jesus' time Jews lived through-out the Roman Empire and were well-regarded in Roman society. It was a time of sacrificial worship and it was quite common for people to visit the temple of their favorite god and have an animal killed as an offering. Later in life Jesus lived in Nazareth, not far from Jerusalem with its great Jewish temple—a temple that was central to Jewish worship and faith in the ancient world.

The publicans written about in the bible were money lenders who got a bad reputation for extorting money over and above the required amount before the would release the debtor from the contract. The Pharisees were experts in Jewish religious laws —a form of religious lawyer evidently held in the same high opinion as lawyers are today.

In the time of Christ. Jewish intellectual life was dominated by the teachings of the great Jewish Pharisee, Hillel, who was really a philosopher and taught the innate worth of man. Hillel made a remark when asked about the meaning of the religious law, which became famous in his time. He said, "The law may be interpreted as do *not* unto other as you would *not* have them do unto you. All the rest is commentary…" Hillel saw religious law as inhibition. Jesus changed this saying into, "Do unto others as you would have them do unto you" and in doing so turned religious belief into a call for action. Christianity started out as a separate sect of Judaism and within a few decades it evolved in a distinctly different form of religion.

Another thing that was popular in Jewish circles in Jesus' time was the Jewish book of mysticism, the Kabbalah. This book tried to promote social harmony by developing a system of social ethic—of seeing things from other people's point of view. This is where Jesus Golden Rule "Love thy neighbor as thy love thyself", comes from. In a very real sense Jesus was trying to become the Jews' Messiah and bring about a new society of this social harmony. From the Lord's prayer "his kingdom his will be done on earth as it is in heaven". Jesus began his ministry by reading a prophesy about the messiah from a Jewish religious text and saying "You have seen this prophesy fulfilled before your eyes". This is why every Christian sect insists Jesus was the Messiah. An idea the Jews reject. This is something that has contributed to the animosity between Jews and Christian.

The Jews did not murder Jesus. Jesus more or less caused his own death. Jesus' death was certain from the moment he over turned the money changers tables in the temple, accusing them of being thieves. This was a direct attack on the corrupt temple priests' source of money. The priests arranged for the death of Jesus out of greed and to protect their power. After his arrest Jesus went mute before Pontius Pilate. Jesus was a very eloquent man who could have talked his way out of any danger by pointing to some of his acts to prove he wasn't anti-Roman—acts like healing the Roman Centurion's servant and telling his followers to pay the Roman taxes (Render unto Caesar what is Caesar's) Instead Jesus kept silent, saying nothing.

Pilate found no error in him and sent him back to the local authorities. The temple priests brought Jesus back to Pilate a second time—this time with the charge Jesus was calling himself king of the Jews. Pilate was still reluctant to kill Jesus and offered the people a choice of between a common murderer Barabbas or Jesus, saying he would pardon one and crucify the other. The temple priests paid the crowd to cheer for Barabbas. So Barabbas was set free and Jesus was crucified and Pilate washed his hands of any responsibility and got the hell out of town.

The Druze are a separate sect of Islam that has secret writings about Christ. In 1985 I was told by a Druze, that Jesus made the temple priests promise that if he was resurrected from the dead, they would declare him king of the Jews and that by crucifying Jesus as the king of the Jews they were mocking him. Like I said I heard that by word of mouth, but it rings true.

Something interesting about the crucifix, religious scholars unanimously agree Jesus was crucified through the wrists and that all Crucifixes that show him with the spikes through his hands are wrong. The reason that Christ is shown with the spikes through his hands is so that the image will conform with scripture. According to the gospels the apostle Thomas doubted the resurrection and said he would not believe Christ had risen from the dead until he met Christ in person and put his fingers through the holes in Christ's hands. All criminals crucified by the Romans were crucified through the wrists. If you attempt to crucify someone with the spikes through the hands the flesh will tear and the person will fall off the cross. The flesh of the hands is simply not strong enough to hold the weight of a man. The church has never seen fit to change the false image of the crucifix.

The church also has not seen fit to publicly denounce the shroud of Turin, which they know to be a fake. The shroud of Turin is supposedly the burial cloth of Jesus. In reality it is a print made on etching, which was then painted and baked in a sliver box. The reason it was baked is because that is the way you age art forgeries. The baking of the shroud affected the paint and the current image on the shroud is the result of chemical burns to the fabric of shroud caused by the heated paint. The burns are not flash burns caused by the resurrection of Jesus as some contend.

They have analyzed the blood stains on the Shroud of Turin and they are paint pigment. On the Shroud, the blood is shown as running straight down Jesus' back and arms in rivulets exactly as if he were hanging on the cross and gravity were pulling it directly down. The problems with that are; blood congeals, pools, follows the curvature of the skin and muscles, and blotches when pressed into cloth. It is the blood which was proven to be Venetian Red paint, by the paint pigment, that gives the shroud away as an art forgery. In the Vatican library there is a letter from the first Church official to investigate the Shroud. In this letter the official plainly states he found and talked to the artist who created it. You must understand the shroud is from the dark ages when religion was an industry and every church had wood supposedly from the original cross and mother's milk from the Virgin Mary.

Another fake the church has never denounced is a silver receptacle, which supposedly holds the foreskin of Christ, saved from when he was circumcised. The Catholic Church's attitude to all of this is "What is the harm if it inspires faith?" The church sometimes has the ethics of a used car salesman. Anything to make the sale.

The resurrection was a matter of dispute in the early days of Christianity. All most all the disciples wrote Gospels. These other Gospels not found in any current bible are collectively referred to as Gnostic Gospels or Gnostic text. (Gnostic is pronounced 'nostic', the G is silent.) The word gnostic itself is where the word agnostic comes from. An agnostic is someone who is not sure there is a God. The Gnostics doubted that Jesus was God and the word agnostic comes from the practice of referring to someone unsure of Jesus as God as 'a gnostic'. There is even a Gospel by Mary Magdalene in which she bluntly refers to Jesus as her husband, implying they had sexual relations. The Gnostic Gospels contended that the resurrection was of the spirit and not the flesh and that Jesus was born of natural means.

The contention Jesus was born of natural means and resurrection of the spirit not the flesh is supported by the discovery of the tomb of Christ. This tomb is the tomb discovered in Israel in the 1970s with five bone boxes in it. These bone boxes were engraved Joseph, Mary, Jesus, Marameme and Judah. Marameme being Jesus' dead wife and Judah his dead son. Marameme should not be confused with Mary Magdalene, the prostitute who was a later day follower of Christ.

This tomb which was publicized in 2006 by James Cameroon, the Hollywood personality, is dismissed by organized Christianity, because it clashes with their contention of the bodily resurrection and bodily assumption of Jesus into heaven. Organized religion contends the names were quite common in ancient times and that it the tomb of some other Jesus not Christ. The odds that this is some other Jesus, who is the son of parents Mary and Joseph are less than one hundredth of one percent.

Do the math, if you figure there are at least two dozen common names for each sex. That makes a marriage of Mary and Joseph one out of 576 possible combinations. The chances a Mary and Joseph would have a son named Jesus one out of 13,824 possibilities and those odds are predicated on the belief there are just twenty-four common names for each sex.

Does that sound like a lot? In reality 24 names is less than one name starting with each letter in the Jewish alphabet. If you say there is an average of four common names for each letter in the alphabet the odds exceed a million to one that the tomb is the tomb of Christ. Something that is lost in the fog this debate is if the tomb is indeed the tomb of Christ, then it is also the tomb of his mother the Virgin Mary.

I would like to say something about the Virgin Mary. There is an excellent reason to believe the Virgin Mary was black. The story I heard by word of mouth was that a Polish artist painted a portrait of the Virgin Mary as a white woman and one night after he was finished the portrait inexplicably turned black as he slept. It could not be a fungus, because only the flesh tones turned black. Some people believe it was a miracle and a sign the Virgin Mary was black. I believe this story when told to me and have since done a little more fact checking.

What is on the internet about this Polish painting of the black Madonna has changed within the last ten years and is interesting. Now Wikipedia says the image was originally painted on a coffee table? That the painting is one of the oldest known likenesses of the Virgin Mary. That this painting has a long history dating back well over a thousand of years. I am a white man but I truly believe the Virgin Mary may have been black. Given the age of the painting and the fact it is the oldest known painting of the Virgin Mary it maybe her actual likeness. It is possible. I don't know if Jesus was black. Joseph was almost certainly Jewish. By far and away the overwhelming majority of Jews are white. This image of Joseph and the Virgin Mary has a distinctive emotional appeal to me. It would make Jesus the product of mixed marriage, meaning he belongs to all races.

The tomb of Jesus raises an interesting question. Did God give Jesus a normal life, before stripping it all away and turning him into God's exclusive servant? Was it the death of his wife and son that drove Jesus out into the wilderness where, he eventually found his calling? The fact that there is only one child who did not survive Jesus, is not surprising. Child mortality was high in Jesus' time and Jesus was a moral man, who probably would not have gone off into the wilderness if he had other mouths to feed at home.

The tomb also supports the idea that Joseph was Jesus' father. Several gospels both common and Gnostic say that Joseph and Mary were engaged to be married, when Mary became pregnant and Joseph married her anyway. I am inclined to believe that story and that Jesus was Joseph's son as the result of sexual relations Joseph and Mary had while engaged. Fully over 90 percent of engaged couples try sex before marriage and it would take a very persuasive angel to convince a man to marry a woman who got pregnant by somebody else, while he was engaged to her when the child could not be his. The origin of the story of the Virgin birth or Immaculate Conception comes from scripture in the Old Testament, "A virgin shall have his child and it will be seen as a sign of the times".

Early Christians embellished the story of Jesus to make it fit Jewish expectations of the Messiah and to better compete with the classic

Roman and Greek gods of the times. Jesus was said to walk on water because the Greek and Roman gods of the sea Neptune and Poseidon could. The lineage of Jesus in the New Testament is in four groups of fourteen generations. Why fourteen, because the Greeks commonly stated the lineage of humanity as 14 generations from the creation of the earth by the Gods. Jesus's lineage would be more convincing if he was descended from Samuel or Nathan, two Jewish prophets who live in the time of King David. Also, the ancient Jews followed lineage from the mother so the biblical record of Jesus' linage could not be accurate.

Look at the times and common assumptions of the people both then and now. Things like God created man in in his own image. Do you believe God, the spirit, in his nature state has human form? The Greek and Roman gods were of human form and had children. Look at the beliefs of the times and you will see the roots of Jesus as the son of God. Jesus as God in human form, not a human being as the Gnostics contended.

To me personally Jesus is nobler and more impressive as a man, who ate, slept, defecated, made mistakes, who had a wife and son, losing them both before he found his calling and preached a doctrine of brotherhood and forgiveness—a doctrine that was certified by God anointing Jesus with the power to heal. A Christ who befriended and redeemed a prostitute, Mary Magdalen showing her she was still worthy of love. Nobler and more impressive as a man, than as a sterile Son of God who never had sex or made mistakes, could walk on water and heal or perform miracles with all the ease of breathing in and breathing out.

Life is about struggle to overcome the many obstacles we all face in day to day life. To me Jesus was a man who did the best he could with the cards God dealt him who when tasked by God, created a legacy which is still dynamic today and is helping to shape the destiny of the human race. I find the image of Jesus as God play acting in a role toward a fixed outcome, he knew all along as sterile and devoid of compelling conflicts and facts that make us want to pay attention to his story and live by his doctrine and example. Make no mistake about it, I am a Gnostic Christian. The belief Jesus was human, not the divine son of God born of immaculate conception, is what Gnosticism is all about.

Three hundred years after the death of Christ the more radical Christian theologians convinced the Emperor Constantine of their image of Christ as the Son of God and the Gnostics who believed Jesus was human were brutally suppressed, murdered and their writings burned. The Gnostic Gospels were lost for sixteen hundred years, until a jar containing a copy of them was discovered in 1946 in Nag Hammadi Egypt. The Gnostic writings recovered include a copy of the sayings of Christ. Some of the more interesting sayings of Christ are "The kingdom of God is inside of you and all around you." Others "I need followers who are clever as foxes but innocent as lambs." and "The truth shall set them free."

The most accurate and authoritarian book on the Gnostic Gospels is entitled simply *The Nag Hammadi Text* which is a literal translation and even has gaps in it where the Gnostic scriptures were damaged and it is not known what was written. This book is currently out of print. One should be wary of books claiming to be copies of Gnostic writings, as they are more often than not *interpreted*—slanted to the author's personal point of view. One of the worst examples of this sort of thing is a book called "The Nag Hamadi Library". It is a sad fact of human nature that a lot of people are fond of bending the facts to support their personal beliefs.

There is another fraud related to the attack on Gnostic beliefs, the so-called Gospel of Judas. Judas hung himself shortly after the crucifixion The Gospel of Judas is a fake intended to discredit gnostic beliefs by saying it is the doctrines of demons. A biblical scholar supposedly found the Gospel of Judas and gave to other scholars. When their interpretation was different than his, he made an amazing discovery of some new portion he had overlooked before and this changed the accepted interpretation of the Gospel of Judas to more along the lines of what he was contending.

The first bible was written in Greek and the British Museum has a copy this first Bible. This bible was translated into the King James Bible. The original Greek Bible was edited from four of the gospels and any controversial passage where deleted. About fifty years after the original Greek bible was written, the Roman Catholic Church rewrote the bible in Latin and left the controversial passages in. An example of this is Matthew 28 of the Catholic Bible, a report of the guards of Christ's tomb being bribed to say the disciples stole the body. This evidently was a part of the original Gospel but is not in the original Greek Bible. The Roman Catholic Church also included other books in the Old Testament, which they thought were relevant.

The men who translated the original Greek bible into the King James Bible were guilty of misogyny and changed some passages to demean women. They specifically removed the verse from the book of Ecclesiastes "Be content in a wife who loves you." They also used a questionable translation of Revelations because the original Greek version was in an early Greek dialect of modern Greek, which is difficult to translate. A more accurate translation of Revelations can be found in "The American National Standard Bible, a New Translation", and was in the Jehovah Witness Bible "A New World Translation" which was changed in 2015 to the more accepted version. There are major differences in Revelations in these bibles and King James or Catholic versions of revelations.

Most of the Gospels, Gnostic or common do not comment on the meaning of the Crucifixion. In Jesus' time it was common Jewish practice to go to the temple and make a sin offering. Have a bird or other animal sacrificed in atonement for some wrong act or misdeed. For three hundred years after the death of Christ, the Christians were brutally persecuted by the Romans because all Romans were expected to pray to the Emperor as a God and the Christians refused. When the Christianity finally won acceptance by converting the emperor there was an effort to develop a coherent doctrine about Christ's ministry and crucifixion. All the Christian leaders met in conference at Niece and what emerged was the doctrine that Jesus was God's only begotten son crucified as God's sin offering and bodily resurrected and assumed into heaven.

The doctrine being that Christ's crucifixion was God's sin offering to remove the stain of sin on all humanity from Adam and Eve in garden of Eden. That in Christ's suffering, death and resurrection there was a new creation, a new beginning for all mankind. The resurrection and bodily assumption of Christ into heaven was the proof of the victory over death and eternal life for all who believe in Christ and followed the teachings and doctrine Christ.

It should be noted that the idea that all people who accept Christ as their lord and savior are automatically saved was not a part of early Christian doctrine, which emphasized good works according to Christ's doctrine and the acceptance of Jesus as God. While the idea Jesus saves is as old Christianity itself. The doctrine of salvation by election was a product of the reformation over a thousand years after Christ. This doctrine holds that all a person has to do to saved is accept Jesus as your Lord and Savior. The man who created this doctrine, John Calvin, burned people at the stake. As Jesus said in the sermon on the mount 'Blessed are you when you are persecuted in my name's sake" That not only includes people who were persecuted for believing in Christ, but also people persecuted by the merchants of self-righteous doctrines in his name.

The reformation itself was a backlash against the corruption of the Catholic church. Every church had wood supposedly from the original cross and mother's milk from the Virgin Mary. If you wanted to be forgiven your sins and go straight to heaven all you had to do was give the church money and you would be forgiven. It was this practice of selling penitence that Martin Luther specifically attacked in the documents he nailed to the church door in Germany. The Church was quite corrupt. Everybody knew it, so Martin Luther's dissidence and criticism fell on fertile ground.

The reformation tore Europe apart, religious strife and persecution became quite common in Europe. The diversification of Christian denominations started before the reformation, but reformation led to a wide proliferation of Christian sects which continues today. Dozens if not hundreds of major churches have been started since the reformation. Growth in existing faiths goes hand in hand with the foundation of new versions of Christian faith in the modern world and no clear sign of the consolidation of the Christian community is on the horizon.

The idea that all you have to do be saved is accept Jesus Christ as your Lord and Savior, that God will forgive all your sins and love you, is very emotional. It is the adult equivalent of a woman reassuring her child that "mother loves you". It is very powerful and salvation can lift people out of addiction and transform people's behavior, providing direct and immediate benefits, but it also commonly abused, by people wishing to avoid consequences of their own actions.

It is quite common for murderers on death row to give their souls to Christ and insist they are saved. O.J. Simpson after his acquittal on charges of murdering his ex-wife went to church and got saved. He in all probably murdered his wife, but now according to his and lot of other people beliefs he has a free pass at the judgment. According to his beliefs he will not be judged when he dies, he will automatically go to heaven. Do you understand the arrogance of that point of view? It is the same sort of arrogance that led O.J to remark "If I did it, she deserved it." According to a great many Christians they will not be judged when they die, they will automatically go to heaven. It is one thing to believe God is a loving and forgiving God, it is quite another thing to think you are not accountable for your actions.

People who believe that in giving their soul to Christ they are automatically going to heaven are often very quick to judge others and commonly insist that only the people that believe in Christ are going to heaven. To them the truth is not what you do, it is what you believe. The idea that the truth is what people believe is the motive for every lie ever told. Propaganda is predicated on that idea. People who think they are automatically saved by giving their soul to Christ are telling God what to do. They insist God's love is unconditional. That no matter what you have done God will let you into heaven if you just accept Jesus as your savior. Ask a born again Christian if God can lie. They will almost always say no. God can lie to you to test your faith. God can put falsehood in prophecy to trap those who try to make prophesy come true to their own liking. God is beyond all human concepts of Good and Evil. While God may be forgiving, God judges, God damns and the idea that you are accountable for your actions is a basic cornerstone of moral behavior.

Jesus was the only prophet of God not to die a natural death or heal. Jesus' ability to heal was a distinct and new thing in human history. Every medically certified miracle cries out that Jesus healed. To say that a healing miracle is indeed possible reinforces the proposition that Jesus healed. God granted Jesus the power to heal to certify his doctrine. All healing miracles can be asserted to be proof that Jesus healed. But in talking to some modern Christians you will find a ready acceptance to say that Jesus suffered for them and their sins and a reluctance to say or admit that Jesus actually healed. They accept the emotional appeal of salvation, but the idea that Jesus healed comes off like a platitude. People will grant that Jesus died for their sins, but don't believe that miracles actually happen. You may not believe Jesus is the Divine Son of God who died for your sins, but the healing miracles of Christ were truly divine in nature and validation of Christ's doctrine and ministry.

This doctrine of Christ's crucifixion as God's sin offering, the idea that Christ's suffering was the new creation, is predicated on belief in the Garden of Eden. If you don't believe in the Garden of Eden, you cut the legs out from under this doctrine. This is at the very crux of the conflict between evolution and religion. The creationists reject evolution for emotional reasons. To say that the crucifixion was not God's sin offering is one thing, but the idea that Christ's ministry and death was not a new creation in mankind's relationship to God is not correct either. In a very real sense, the cross of Christ is the essential foundation of Christianity, the suffering of Christ and his followers lead to a new birth in faith.

There is a phenomenon which can be considered proof that Jesus' suffering was divine in nature: it is known as stigmata. This is where a person becomes afflicted with the wounds of Christ. Jesus had five wounds from his suffering or his passion as it is known. They are: One, the wounds on his forehead and scalp from the crown of thorns. Two, The whip marks on his back. Three, the holes in his wrists from the crucifixion, Four, the holes in his feet also from the Crucifixion. Five, the spear wound in his chest from the Roman guard who made sure he was dead. Nobody has ever manifested more than three of the wounds. It is a very well documented and unexplainable phenomenon. The shroud of Turin is a fake, an artistic forgery, but stigmata manifests itself in different people over time. In a real very sense it is evidence of the divine nature of Jesus' suffering that is new and current every time it manifests itself.

So, what is one to make of the crucifixion? Jesus went mute before Pilate and accepted his fate as a duty to God. In the sermon on the mount Jesus said, "Blessed are you when you are persecuted for my name's sake". All the apostles suffered for their beliefs, some died horribly. Thousands upon thousands if not millions of people have died to make Christianity what it is today and Jesus shared in their suffering. Indeed, Jesus went first and before he suffered Jesus shared of his flesh and blood in the wine and the bread of the last supper. Jesus suffered and offered others a way to share in his suffering. All people suffer illness, the loss of loved ones, and sometimes as victims of crimes or injustice under the law, but very few people suffer as much as Jesus did. Communion, the bread and wine, the symbolic body and blood of Christ, is a covenant with Christ to accept your suffering in life and live to the higher ideals of love, tolerance and forgiveness and that Covenant with Christ is at the very heart and soul of Christianity.

ISLAM

"All is Vanity"
a phrase used repeatedly in the biblical book "Ecclesiastes"
and a frequent saying of the prophet Mohammed.

What a lot of Christians don't understand about Islam is that Muslims believe that Jesus Christ was a true prophet of God who did heal. Most Americans don't understand Islam, but in fact Islam is one of the fastest growing religions in the United States. It is not unusual to see women in public with their hair covered. They are not bothered or sexually harassed. Men commonly know that such a woman is not interested in being picked up and their attention is not wanted.

There was a Muslim man who was elected to Congress. He took his Congressional oath of office with his hand on a copy the Koran. There are Muslims in the armed services who have served in Iraq and Afghanistan. The Army has a Muslim Chaplain who asked for and received from, the National Muslim Council, a Fatwa or religious decree to fight the war on terror. The mastermind of the 9/11 attacks at his trial before a military tribunal cited the wars in Iraq and Afghanistan as proof that America was attacking Islam. Neither war existed before 9/11 attacks on the United States. If there is a war between the United

States and Islam, he and Osama bin Laden started it.

Islam literally means submission and was founded by the prophet Mohammed over 1400 years ago. Mohammed began to have strange dreams. At first, he thought he was going crazy, but then accepted the dreams were in fact divine in nature and had scribes write them down. These dreams became the Koran. All Korans are exactly the same and the Muslims believe it is a sin to translate the Koran.

At the time these dreams began the prophet Mohammad lived in Mecca which was a center of pagan worship. His ideas were not well received and he and his followers were driven out of Mecca and the Muslim Calendar begins from the date of this flight. He and his followers took up residence in Medina where Mohammed continued to build Islam. He eventually raised an army and conquered Mecca and made it the center of worship for Islam. All Muslims pray toward Mecca and only Muslims are allowed to enter Mecca.

Muslim prayer is a physical act of submission to God. You start on your feet bow then fall to your knees and press your head into the ground. You then repeat the last step several times. You are literally humbling yourself before God. It is hard to understand how complete this attitude of humility is in devout Muslims. I lived and worked in Saudi Arabia from the summer of 1984 until the spring of 1986. One night when driving back to the compound in Riyadh with a couple of friends we saw a fire in the distance and went to check it out. Some of the poorer people had built a series of shacks out of building material, which caught fire. We stayed until the fire had burned itself out, then returned to our compound.

I couldn't sleep that night and returned to the site of the fire. Despite the fire all men in this community gathered together to pray at dawn. They were led by an elder who by the teaching of Mohammed held two strings in his hand, one white, one black. At dawn when he could tell the difference between the two, he led the men in prayer. As I stood there watching them the hair on the back of my neck stood up. There have been a couple of times in my life when, I have been moved by the beauty and the majesty of worshiping God. That was one of those times.

For a few hundred years after its creation Islam flowered. When Europe lived in the dark ages Islam led the way in medicine, science and mathematics. Now Islam is often attacked as a medieval religion because murderers are beheaded, thieves have their hands cut off and adulterers are stoned. Islamic law is based on Ten Commandments. While this strict code may seem to be inflicting cruel and unusual punishment, it virtually eliminates crime.

It normally takes four male witnesses or eight female witnesses or some combination of the two testifying they saw the actual physical act of penetration to get a conviction for adultery in an Islamic court and have someone stoned to death. Most of the deaths associated with adultery in Islamic countries are actually honor killings. The only execution for adultery in my eighteen months in Saudi Arabia was of a man who molested a small boy. The court accepted the testimony of the boy and ruled the crime adultery because the man was married, so the man was put to death.

In the time I was in Saudi Arabia there were less murders in that country of five million than there were in my hometown of Richmond, Virginia—a city of 250,000 people. It wasn't even close. The numbers were dozens to hundreds.

In the time I was in Saudi Arabia, no thieves had their hand cut off. A thief only has his hand cut off on the third offense. Thieves do exist in Saudi Arabia. Saudi Arabia imports a lot of third world laborers who steal. Merchants in the big cities lock up when they leave their shops, but in small towns where there is no third world labor, it is quite common for a merchant to leave his store unlocked and unattended when he goes off to pray. As late as the forties and fifties travelers reported seeing gold bars left unguarded and unattended in Saudi airports. Gold bars mind you, try leaving a boom box unguarded on some city bench and see if it will be there in an hour? I doubt it will be. So, don't be so fast to judge.

Islam has over time become abusive of its women. In the days of Mohammed women went unveiled and led armies. Part of the problem today is due to one of the sayings of the prophet Mohammed being taken out of context. There are thousands of sayings of the prophet Mohammed which were written down and are considered by some to be holy. The saying I am talking about is "Women are as tempting as the devil himself" This has come to be interpreted as proof that woman are evil and by nature sexually immoral. I am convinced that this remark is taken out of context. In a private moment whisper in woman's ear, 'You tempt me". She will take it as sexual flattery. Mohammed could have easily used his remark to greet large groups of women and they would have taken it as a form of flattery. It would have made them smile. At this point does anyone know the context in which Mohammed made his remark?

Another reason women are seen as immoral is the ease of divorce. As Jesus observed when you divorce a woman you make an adulteress out of her. When a man divorces a woman who bore him children, the divorced woman will screw around to punish the man who abandoned her: it is not rational behavior but it is a distinctive feature of a woman's nature. Perhaps if Islam were to make it harder for a man to use a woman and throw her away, they would find their woman will be more moral.

Women are by custom excluded from prayer. Women do not enter mosques because they are thought to be unclean when menstruating and if they attended services on a regular basis men could tell when they are menstruating by their absence. In Leviticus, the Bible says women are unclean for ten days after they menstruate. If you never had sex with an unclean woman, you have never fathered a child, because women ovulate four or five days after their menstrual period and are no longer fertile by the tenth day.

Some of the passages in the Koran are disturbing to women. The one with which I am familiar is: "If a woman will not have sex with her husband she should be raped." No self-respecting woman accepts that as the divine word of God. This passage and others maybe the cause of the Satanic Verses controversy in Islam. I am not going to comment on that controversy. I will however say the following, which I truly and firmly believe. The prophet Mohammed, blessed be he and his name, was a man, not God. When you say Mohammed made no mistakes, that all his teachings and saying are perfect you are turning him into a God for your own personal reasons. What you are doing is contrary to the teaching and spirit of Islam and you are committing blasphemy. The state of relations between men and women in Islam is poor and misogyny is very common. Women agitating for a more equitable deal is part of the reason why some Muslims see Islamic culture as under attack.

Some uninformed people make fun of the Hajj, the annual Muslim pilgrimage to Mecca. People feel moved when they attend a major sporting event, where the crowds range from 50 to 100 thousand. Others speak fondly of the peace and love of the Woodstock music festival attended by over a quarter of a million people. The Hajj regularly draws two or three million. A veritable sea of humanity and there is something intrinsically exciting and awe inspiring in being a part of such a mass of humanity. You meet Muslims of different ethnic origins.

Share bread and food with people who were total strangers before you met them. The Hajj is a moving experience that reinforces the teachings of the prophet Mohammed. The Hajj is how one becomes born again in Muslim faith.

I want to say something about a basic tenet of faith of Islam. The common saying: "There is nothing in this world but the will of Allah." I have a devout Muslim friend named Ali Al Marsi. When I spoke to him on the phone for the first time in years over two decades ago, he informed me he had three sons. I teased him about this, as it is a sign of virility in the Muslim world. His response was humble: "Whatever sex God chooses." Ali thought God deliberately picked the sex of his three children. He made no allowance for chance in the choice.

I was in a discussion about religion and was treated to a different interpretation of this saying. I was discussing religion with an educated Muslim friend when I asked him, "Do you really think God controls every little act, every word spoken, every little event in the world or that God allows things to happen, including bad things to happen?" to which my Muslim friend replied "Even if it happened by chance, it is still the will of God. Because Allah allowed it to happen." That is an argument I can accept.

Because of this idea God controls everything. Muslims are much quicker to take offense, if you speak ill of or slight a Muslim in the Middle East, those Muslims will interpret that as a sign you choose to be their enemy. There is no such thing as a good-natured ribbing in the Muslim world. Muslims are also inclined to see things that severely impact them as evil. This is part religion, part culture, part ego. But it is a fact of life in many, many Muslim countries.

The founder of Isis was in an American prison for a number of years. Instead of accepting this as the breaks in life, he took grave offense and

Of Science and God

plotted the destruction of the society who visited this injustice upon him. He didn't object to the injustice. He took it very personally. He developed an ideology of blanket condemnation of western society and set about building a society which was based on a strict Islamic interpretation of Islamic law. He then set about building this society while attacking the West. This is an extreme example of ego and slight. But it is to some extent true of all Muslims in the Middle East. There is no such thing as an accidental slight to a Muslim. Deliberately slight them after the initial meeting and you are choosing to be their enemy. A good rule when dealing with Muslims is, "Leave Muslims alone and Muslims will leave you alone." Ignore this rule at your own peril.

The Muslim religion does not have the same Christian ideas of heaven and hell. There is no clear doctrine of eternal damnation: there are levels of punishment and it is possible to be released after a period of time. To some extent, you either make it into paradise or don't there is no strict divide between heaven and hell. During the hajj, one of the rituals that is performed is symbolic stoning of the devil in a valley between Jeddah and Mecca. Satan is not seen as the ruler of hell but rather as the corrupter of society and the social order established by the prophet Mohammed. Life is not a struggle for salvation but rather a quest for a tranquil existence. There is an Arab curse "May you live in interesting times" Muslims seek to live their lives as serenely as possible and disturbers of the established order are seen as trouble makers. There is no quest for salvation in Islam. Life is its own reward and Muslims are not so concerned with avoiding damnation as they are with living life without needless hassles or drama.

Muslims mark the celebrations of their religion with a lunar calendar. The earth and the moon evolved in separate sections of the solar systems. The moon is thought to have started out as a moon of Saturn when it escaped the pull of its planet and began to orbit the sun independently. Eventually the earth and the moon collided which is how the earth got its spin and tilted axis. The seasons are a result of the moon. We have day and night because of the Moon. The lunar calendar can be argued to be a celebration of the rhythms of life here on earth. It also can be used to explain human evolution.

They have literally located hundreds of planets in the nearby stars of our galaxy and not one of these planets rotates. One side of the planet is constantly facing the star that serves as their sun and all the creatures live in constant sun light while the creatures on other side of the planet live in constant darkness. So, in terms evolution stories little green men who draw nourishment from the constant light of their sun like the plants do here on earth are not as absurd as they sound. After all, the human skin uses the sun light to produce Vitamin D. It makes sense that other creatures who live in constant sunlight would have evolved with skins that produced sugars necessary for their metabolism. Likewise, the idea of creatures with large eyes and luminous skin, so they can see each other in worlds of constant darkness makes evolutionary sense as well. Humans are creatures of day and night and the odds that there are other creatures like us exceed millions, perhaps even billions, to one. The earth and humanity will still be here millions of years from now and I think God, in Islam's lunar calendar as established by the prophet Mohammed wants humanity to mark the passage of that time according the instrument of our unique evolution, the moon.

Islamic fundamentalists believe Islam is under attack from the west. In reality it is the free flow of ideas in an increasing smaller and smaller world, which is challenging Islam. Darwin's theory of evolution is a bigger challenge to Islam than it is to Christianity. Adam of the Garden of Eden is thought to be one of five prophets of Islam, Abraham, Moses, Jesus and Mohammed are the other four. Most Muslims believe in the biblical story of the Garden of Eden and have never come to terms with evolution.

Of Science and God

The prophet Mohammed was human and as such was, as all men are, imperfect and in declaring Adam a prophet of Islam, he was wrong. There was no Garden of Eden, no Adam. Where in the Koran does it say the first man and woman were created in the Garden of Eden? If you say Mohammed was perfect that he made no mistakes in his teachings, you are saying he is God and that is blasphemy in Islam

We are all God's children in that we define our own self-worth in relationship to God. This is true even if you personally reject God and seek your value as a person in terms of human secularism. In large part the Muslim Jihadist movement has the same emotional roots as Christian creationism. A backlash against the erosion of religious doctrine as it clashes with the tenets of science. This will continue until Islam comes to terms with evolution and accepts that the prophet Mohammed was a man of his times, that in declaring Adam of the Garden of Eden a prophet of God, Mohammed was wrong.

Again, I repeat, if you say Mohammed made no mistakes-that all of his teachings were perfect, you are saying he was God and that is blasphemy in Islam. There was no Adam or Garden of Eden and until there is widespread acceptance of evolution within Islam, the more radical elements of Islam will continue to see Western technological society as a corrupting influence and see it as attacking Islam and respond with the call for Jihad as a means for defending Islam against Godless Western society. The Nigerian Jihadist even named their movement "Boko Harem" which when literally translated means "No Education." Billions of people all over the world get their own sense of self-worth from their religious beliefs. Try to take away a child's security blanket and the child will resist. It doesn't matter if the Child is four or forty. It is basic human nature.

Judaism

"Our elder brothers in faith"
A term Pope John Paul II used to refer to the Jews.

Judaism is one of smaller religions in the world accounting for far less than one percent of the world's population. It is however the oldest monotheistic religion on earth. Considering anti-Semitism one can say the Jews sometimes inspire hatred all out of proportion to their numbers. Three of the five prophets in Islam are Jewish, Abraham, Moses and Jesus. So, Muslims who hate Jews do so for vain and personal reasons not out of some religious doctrine that compels them to do so. The Jewish calendar dates back to the covenant between Abraham and God that made them a nation.

Judaism is truly diverse with separate sects proliferating among the body of followers. In today's modern world there is such a thing as Jewish Buddhists. An interesting thing, as Jesus who was a Jew meditated according to the Gnostic Gospels. Jesus was a Jew and grew up in the more cosmopolitan atmosphere of Alexandria, Egypt. There was a second temple there where Jews made sacrifices unto the Lord. Ancient Judaism was based in Israel, but the Jews were by immigration and forced capture were spread all over the ancient world.

The temple in Jerusalem was the center of worship and Jewish cultural life in pre-Christian era Judaism. Jews from all over the Roman world sent gold to the temple in Jerusalem. Because of their common language, Hebrew, the Jews have always had a strong national identity. From the time of Roman conquest until about forty years after the death of Christ the Jews of Israel agitated for independence from Rome. There was a group of Jewish terrorists known as the Scari who used to kill Roman sympathizers in crowds. After nearly a hundred years of agitation and forty years after the death of Jesus the Scari succeeded in bringing about an open revolt in Israel against Rome. Rome responded by destroying the Jewish temples in Alexandria and Jerusalem. Rome sought cultural eradication of the ancient Jewish religious practices.

What grew out of the aftermath of this first holocaust was Rabbinical Judaism. Rabbi was the Hebrew word for teacher. Under Jewish laws and traditions, if there are more than ten children in a community, the community appoints a teacher to teach them Hebrew and the Torah. For thousands of years the Jews have kept alive their language, religion and distinctive identity. Orthodox Jews still refer to their places of worship as Synagogues and follow traditions handed down and followed for over a thousand years. Reformed Jews are more modern and blend in with cosmopolitan habits and call their places of worship temples. This signifies they are a faith not tied to the dogmatic interpretations of the Jewish faith as a necessary part of worshiping of the Lord. Conservative Jews fall more in the middle between these two opposite facets of Judaism.

All Jews regardless of their sect read the Torah during services. The Torah consists of the first five books of the Bible. These five books are the oral history of the Jewish people and some of the laws by which Jews lived, which were written down and converted to a written record when the Jews invented the Hebrew alphabet and their written language. Some of these scrolls are literally thousands of years old. Like the Koran, all copies of the Torah are the same. One reason the members of Judaism have prospered economically through the ages is it places a great store on education. Literacy in Hebrew is a requirement of the faith and getting an education is seen as a duty.

Most people are familiar with the story of Joseph and Moses and the Jews flight from bondage in Egypt. The followers of Islam, Christianity and most people in the Western world are familiar with the story of the Jews flight from Egypt and parting of the Red Sea. If you want a different slant on the parting of the Red Sea, read a book entitled *When World's Collide* by Immanuel Velocoski. This man was a scholar who maintained that the disturbances in Egypt which are described in Exodus were the result of the planet Venus passing close by earth due to a cosmic disturbance.

We live on one of the spiral arms of the Milky Way. It is not impossible indeed it is quite likely that from time to time other stars pass close to the earth and sun, causing general havoc in the solar system. Astronomers say there used to be an atmosphere and water on Mars. What happened to them? The earth has most likely been subject to cosmic trauma as well.

It has been 15 thousand years since the last Ice Age. Sea levels drop dramatically during an ice age and the Red Sea would have been much smaller. The Jews could have crossed the Red Sea south of Suez. The close passing of another planet inside the orbit of the moon could have easily produce tidal surges hundreds or a thousand feet high. Moses and the Jews, could have beaten Pharaoh's army to the high ground in between these tidal surges. They would have had about 12 hours to make the crossing and literally would have been running for their lives, both from pharaoh and the wall of water Moses most certainly knew would surely come.

Venus' atmosphere is caustic consisting mostly of sulfuric acid. The clash of the magnetic fields and the motion of the two planets could have easily led to electrical discharges between the two planets. Like the tides the atmosphere would have been in turmoil, cyclones and tornadoes. If the two atmospheres mixed it would have burned to breathe. Something that would have made the flocks and herds, stampede. With man and beast running for their lives under a sky lit by interplanetary lightning and atmospheric lightning, it would be a truly biblical upheaval.

It may be possible to one day identify the spot where Moses crossed the Red Sea. According to Muslim oral history, after Moses and the Jews crossed the Red Sea Moses raised his staff, struck a rock and a spring emerged. This act may sound a little farfetched to the non-believer, but Moses may have been a diviner. Someone who finds water by holding his arms straight out from his chest with rods or sticks in his hands. As a teenager, I saw a 60 Minutes article by Morley Schafer on the subject. 60 Minutes is not in the habit of selling crap and they were duly skeptical. The report was enough to convince me there is something in it.

Velocoski made many predictions about the planet Venus which have proven to be remarkably accurate. The science in the book is not very good but the book itself gives one pause to stop and think. The whole thing is a subject that deserves to be studied scientifically. Computer models could be made of what such a close encounter would do the earth's oceans and atmosphere. Studies of Venus orbit and studies of nearby stars to find the culprit—the star that came close to the solar system and caused the upheaval. Perhaps, a Hebrew university in

Israel would like to sponsor the project, a search for a historical and scientific Moses.

Some Jews reject God because of the Holocaust. It is an egocentric perspective common to questions of religion and faith, the egotistical question-how can God let me suffer? Jews will commonly tell you that Jesus was a Jew and a great man, but they do not think he is the divine Son of God. Discussing Jesus with Jews makes them very uncomfortable. Because they are very conscious of nearly 2,000 years of persecution as the killers of Christ. The Jews in some respects see Jesus as a traitor to Judaism. As a man who abandoned his faith for egotistical reasons. Someone who sought vain and personal glory in the new religion he founded.

Still Jesus has had a major influence on Judaism. Jews commonly teach wisdom by parable a practice Jesus started. They have refined it into an art form and it is an integral part of Rabbinical Judaism. The reformed Jews changed an ancient Jewish prayer that says God can raise from the dead, in a backlash against the resurrection. Jews do not believe that Jesus healed. Look up Jesus in the Jewish encyclopedia and you will find they consider him to be a magician. This is understandable as the Jews did not take a hard look at Christianity until decades after Jesus' death and faith healing was quite common early Christianity. Yet Jews will pray for the recovery of members of their congregation who are ill. Jews bitterly reject the Christian doctrine of Jesus as God. In do so Jews also reject Jesus as a prophet of God who did heal. The animosity between Christianity and Judaism will never be healed until the Jews accept Jesus as a true prophet of God who healed—a Jew called by God to another purpose.

The Jewish attitude toward sin and salvation is interesting. The Jews have no doctrine of salvation in Judaism. There is no sin-mongering.

No calls to repent of your sins. Sin is not seen as what separates you from God. Rather sin is seen as your selfish acts which harm others, and the holiest day of the year is Yom Kippur. The day of Atonement is the time when all Jews humbly ask God to forgive the sins that they committed in the past year. It is a very solemn serious thing to them.

Jews are willing to accept the evolution of humanity. Judaism's roots are deep and the entirety of the religion holds the seeds of a new renaissance of faith sometime in the future. Judaism is the oldest of monotheistic faiths. It has endured many trials and tribulations and remains largely intact. Judaism stands up very well to the challenges of science. Many people who describe themselves as Christian, quite a large percentage of them, essentially believe Jesus was a man born of natural birth and doubt the resurrection, while ignoring the doctrine of Christ. Their viewpoint is very Jewish. If the world is renewed in faith Judaism will prosper. It is a sin to convert a Jew. Argue doctrine with them yes, but don't try to convert them. By asking a Jew to convert, you are asking them to betray their heritage. A heritage they share with Jesus. Jesus was a Jew and there must always be Jews.

The Eastern Religions

"All suffering is due to desire"
a basic tenet of Buddhism

The Eastern religions are marked by a distinct absence of a doctrine of Salvation. They do not have churches where they go to worship one day a week. Temples instead serve as a reminder of those values the religion emphasizes. In Hinduism, there are a variety of gods with temples dedicated to them. Each god manifests some particular virtue or set of virtues to be emulated or flaws to be avoided. The Hindu religion is in many respects like ancient Greek or Roman mythologies, with their pantheon of gods and stories of their interactions. Hindus believe in God and the immortality of the soul. That the soul is born again and again until one reaches a state of grace and goes to be with God in the afterlife. Hinduism is older than Christianity or Islam but is largely confined to the Indian subcontinent though immigration has spread to many parts of the world.

Buddhism originally founded in India has a wide appeal throughout Asia. Buddhist temples are quite common in Asia. Buddhism has no place for the worship of God in its doctrine. Instead, Buddhism emphasizes reaching a state of grace through its philosophical doctrine.

The Eastern religions are for the most part philosophies. These doctrines have a broad appeal and affect society in a diverse and varied manner. The Chinese communist party which has abandoned communist economics in favor of capitalism is in a very real sense doing what is best for their society according to the teachings of Confucius. As modern society and technology break down the culture and cultural roots of these ancient societies the practice of Christianity is growing. The growth of Christianity is slower and less marked on the Indian subcontinent, which has two religions Hinduism and Tibetan Buddhism, which both teach the immortality of the soul and reincarnation. (Though the works of Mother Theresa led to a growth in Catholicism in India.)

Unlike Islam, Christianity and Judaism, the Eastern religions do not believe in or teach the story of the Garden of Eden. So, there is no conflict between their religions and evolution. But mass communications are making the world a smaller place and the religions and spirituality of Asia is evolving just as they are in the rest of the world.

Prayer

*"Do not pray in public like the publicans and
Pharisees, instead pray in a closet"*

the philosophy and teaching of Jesus Christ about prayer.

Praying to the gods or God is as old as civilization and prayer itself is linked and intertwined with religion. Muslim prayer is very strict and a physical act that humbles one. The Jews have a great variety of prayers written down and used during their services. In the Catholic Church prayer is written down and is referred as the liturgy. Outside of Catholicism Christian prayer is by-and-large up to the discretion of the Christian minister conducting the service and is invariably from his or her point of view.

There is often an ego-centric approach to prayer. I knew a woman who had a stroke. When I visited her in the hospital the next morning, she said "You can forget about God, I prayed last night and nothing happened". I asked what she prayed for: she grew irritated and said she prayed to be healed of her stroke. She thought there was no God because he didn't answer her prayer when she prayed with all her might. She should have prayed for strength in the recovery because that is what God gave her. If God answered all prayers, everybody would win the lottery and no one would ever die.

Miracles can be argued to be the result of prayer. People suffer all manner of illnesses and God's intercession to relieve suffering is extreme rare, considering there are billions of people in this world and only a handful of miracles. When God does act it doesn't always make the news. The church has documented many more miracles in the last thirty years than two Catholic miracles that made the news. The miracle which was associated with Mother Theresa was described earlier, another miracle was a remission of a nun's MS after all the other nuns in her convent prayed to the dead Pope John Paul II who also had MS. Miracles only appear to be granted to reward those praying not the prayers of the person suffering.

Most prayer in Christian churches serves the function of indoctrinating the parishioners to the minister's point of view. Prayers are used to shape attitudes. Jesus tried to teach people how to pray, with the Lord's Prayer.

THE LORD'S PRAYER

Our father who are in heaven hallowed be thy name,
thy kingdom come thy will be done on earth as it is in heaven.
Give us this day our daily bread and forgive us our trespasses
as we forgive those who trespass against us.
Lead us not into temptation but deliver us from evil,
for thine is the kingdom and the power and the glory, forever
and ever. Amen

Christ failed miserably in his attempt to teach people how to pray. All manner of prayer is quite common in the Christian denominations. I went to Catholic services on the Pentecost and was appalled to find they had even changed the Lord's Prayer. Most ministers or preachers are guilty of the same sin, using prayer to indoctrinate the faithful to their particular attitude or point of view. Human nature hasn't changed at all since Jesus' times and in effect these ministers or preachers are praying in public just like the publicans or Pharisees. Making a great show of parading their righteousness in public to win approval or convince others of their particular vision of society. "God save us from the sinners", "God save us from this and God save us from that...".

Prayer does have a valid purpose, it can give comfort in times of trouble, lift a heart or restore faith to a person. The effect of prayer in shaping attitudes is undeniable. Faith is ultimately the hope that there is a meaning to life and there is an interesting thing in religion and the nature of prayer. Prayer is more effective when it appeals to hope than when it makes a pronouncement about the certainty of some tenet of faith. For instance, the sailor's burial prayer, "In hope of the resurrection and the Second Coming when the sea shall give up her dead." In the hope, in the hope and there is not one word about Mary's virginity in the Hail Mary prayer. Many people doubt the Immaculate Conception. Not mentioning Mary's virginity in the Hail Mary removes that moment of doubt and frees Catholics to venerate Mary as the mother of Christ. In the Hail Mary prayer Mary is actually referred to as "Holy mother of God". The Hail Mary prayer is unique to Catholicism and helps shape that religion.

Catholics are seldom born-again Christians. The Evangelical movement which places a great weight on salvation through election of Christ as your savior, has powerful emotional roots in the idea. While we are not worthy God loves us anyway and his love for us is manifest in Christ. Why do Catholics venerate Mary? Because the Catholic equivalent of the sentiment, we are not worthy of God's love, is manifest in the last sentence of the Hail Mary prayer "Holy mother of God, pray for us sinners now and in our hour of death". You are literally pleading with the mother of Christ to pray for you and your soul when you die? It is one the most powerful and influential prayers in all religion.

HAIL MARY

> Hail Mary full of grace. The Lord is with thee, blessed art thou among women and blessed is the fruit of your womb Jesus. Holy Mother of God, pray for us sinners now and in our hour of death.

Let us look at some other prayers and how they affect and reflect their religions. First from Judaism. The 23rd Psalm. This is the only psalm that endures. By endures I mean a sizable portion of the population can recite passages from it, even if they cannot recite it as literally written. The psalms used to be poems or songs of praise to the Lord. King David wrote many psalms and performed them by singing them and playing them on musical instruments.

THE 23RD PSALM

> The Lord is my shepherd. I shall not want. He makes me lie down in green pastures; He leads me beside the still waters. He restores my soul. He leads me on paths of righteousness, for his name's sake. Ye thou I walk through the valley of the shadow of death, I will fear no evil. For thou are with me: Thy rod and thy staff, they comfort me. Thou preparest a table before me in the presence of my enemies; you anoint my head with oil; my cup runneth over. Surely goodness and mercy shall follow me all the days of my life. And I shall dwell in the house of the Lord Forever. Amen

The next prayer is Muslim and is loosely translated from the Koran. I came across it in my readings and it is a favorite of mine. Because I believe in an infinite and eternal universal with other forms of intelligent life and in the prayer, Mohammed referred to God as the "Lord of all worlds." Like I said, Islam has the seeds of acceptance of the modern day scientific thought in it. Let us hope it comes to pervade Islam without too much turmoil and bloodshed.

AL FATIHAH

In the name of Allah, the compassionate, the merciful. All praise belongs to Allah. Lord of all Worlds. The Compassionate, the Merciful, Ruler of Judgment Day. It is You we worship and to You we appeal for help. Show us the straightway, the way of those You have graced, not of those on whom is Your wrath, nor those who wander astray.

The next prayer is Hindu and it reflects the common belief in God and in the faults like Jesus' seven deadly sins so common to so many religions. It also reflects the praise of God that runs through so many prayers of so many religions.

HINDU PRAYER

Grant us inner spiritual strength to resist temptations and to control the mind. Free us from egoism, lust, greed, hatred, anger and jealousy. Fill our hearts with divine virtue. Let us behold Thee in all these names and forms. Let us remember Thee. Let us ever sing Thy glories. Let Thy name be ever on our lips. Let us abide in Thee forever and ever.

I would like to include the serenity Prayer from Alcoholics

Anonymous. It is said after ever AA meeting. There is a longer and fuller version of this prayer originally written by Reinhold Niebuhr. But I do not include it here. Instead, I included the short version. I include it in part because it is good advice to anyone facing uncertain and troubling times in their lives.

Serenity Prayer

God grant me the serenity to accept the things I cannot change; the courage to change things I can and the wisdom to know the difference.

The final prayer is one I came across recently. I include it because it reflects the almost universal acceptance of God the creator and the idea that, we go to him when we die. It is an American Indian prayer from the Lakota Sioux Indian Chief Yellow Lark, 1887

Sioux Prayer

Oh, Great Spirit, whose voice I hear in the winds and whose breathe gives life to all the world, hear me. I am small and weak. I need strength and wisdom. I seek strength, not to be superior to my brother, but to fight my greatest enemy – myself. Make me always ready to come to you with clean hands and straight eyes, so when life fades as the fading sunset, my spirit will come to you without shame.

These are some examples of prayer and its power to uplift hearts and fill the soul with a yerning for a better existence in death. The power of prayer maybe abused by preachers or ministers with some particular platform or program they want to sell, but its ability and power to shape attitudes is undeniable. I have thought a lot about criminals.

Michael Lewis White

Some men see businessmen cheat people and governments kill thousands or millions in war and decide it is their right to do the same thing in and of their own accord. Since the beginning of the human race men have fought and abused one another: it is part of human nature. God is certainly aware of human nature and the fallibility of humans and these men are not necessarily damned. Yet there is no prayer for people who engage in crime, so I wrote one. I hope it makes criminals think and be more accepting of God and judgment.

The Criminals Psalm

Accused am I in the community of men. The servants of the law denounce me, the self-righteous castigate me-are we all not flawed. Is not the noble cause the tool of those who thirst for power? Who among us is pure?

Lord you alone know the truth of all souls, am I so low as to be worth nothing, while the high and mighty are to be exalted. Lord, you alone shall judge me. No one, not even me knows the fate of my soul. Lord, why did you make me the way that I am? My life is driven by hard circumstances, can you not find understanding and mercy in your heart for me.

Please, Lord, do not let me be a fool.

The Abuse of Faith

*"If Jesus saves, he better save himself from the gory
glory seekers who use his name in death"*

from a song entitled 'Hymn 43' by the rock band Jethro Tull.

It is a common thing for some people to think they own the truth and specifically includes arguments about the intent, nature or truth of God. The assertion of someone that they understand God's true purpose is a very dangerous form of absolutism. Beware the merchants of moral purity, those who claim to know the truth of God. They are often quick to proclaim this or that or some person is an evil that must be eliminated. This self-styled and self-proclaimed righteousness is often a justification for murder. Hell is positively filled with people who were convinced that they were doing the work of God.

Abortion clinic bombers and shooters are an example of this. These people were commonly abused as children. The innocent they seek to protect are really themselves. Ask a forensic psychiatrist about the man who shot an abortion doctor and his bodyguard in Florida, he was almost certainly badly abused by his mother and was symbolically punishing his father for failing to control his mother. The Massachusetts' abortion clinic shooter who shot women was badly abused by his father and was symbolically punishing his mother for failing to control his father. Children who are badly abused blame themselves for the punishment and the abusing parent becomes untouchable and sacred. Ninety percent of all criminals were abused as children it distorts their moral development.

Men are biologically driven to dominate their environment. All men want to be the leader of the pack. It is quite common for mullahs and Christian preachers to feel they know the truth and preach about what is required of others in the service of that truth. Ministers or mullahs derive a sense of power from whipping up excitement, preaching a sermon which excites and motivates the audience. Preachers and mullahs commonly denounce some group of people to do this. In lecturing from the pulpit, they are speaking about the only morality they think is valid, theirs. How often has a preacher or mullah stood before the people and say 'I was wrong.' In preaching the truth of God, you are expressing an opinion not a fact. It is the biggest touchdown or goal you will ever score, to stand in front of group of people and stir them into a fever pitch. The greater the excitement you produce, the more rewarding it is. There is nothing more subversive than the power of pulpit.

When a preacher or mullah calls for moral purity as a strict observance of faith, they are often practicing a form of bigotry. They are saying only people who behave as they want are worthy of God's love or mercy. Like all bigots and racists, they suffer from a lack of self-respect and by demeaning others they are attempting to bolster their own egos, but only demean themselves. The more secure you are in your own identity and beliefs the less threatened you feel by others who believe or behave differently. It was a Muslim Anwar Sadat, who was so devout he had a bald spot on his forehead from a lifetime of pressing his head into the ground during his daily prayers, who had the courage to make peace with the Jews. That is the kind of behavior true faith promotes.

Compare that to those who denounce others. Where men preach the truth of God and what God wants, you will often find self-righteous, self-serving falsehoods that inspire hate. Jesus said, "Judge not lest ye be judged." Think about that the next time you hear some religious leader denouncing some group of people. All people will be judged and you are not God.

Sex is considered by many religious authorities to be evil or at least sinful. Don't tell anybody but there is a rumor going around that sex is actually necessary for the survival of the human race. All you ministers and mullahs can relax, it is just a rumor. Minister and mullahs also commonly denounce any form of sex other than coitus as sinful. Priests who do this are projecting their own personal struggle with sex and celibacy onto others. Because according to the doctrine of the Catholic Church, any form of sexual contact is permissible as foreplay so long as ejaculation is into the vagina for the purposes of procreation. It is natural and healthy for couples to explore sex. They almost always develop a routine with which that both partners are comfortable.

Many Christian preachers denounce sexual education and preach the teaching of abstinence. Teenage sex has been around for about five million years and simply teaching and preaching abstinence doesn't work. If you don't want your teenagers to have sex you should be an advocate of sexual education. This is because of the fact teenagers are naturally curious about sex and the more ignorant they are about it the faster curiosity kills the cat. Numerous studies have repeatedly shown that extensive sexual education in early adolescence delays the onset of sexual activity in teenagers.

Teaching sex education does not mean you are endorsing sexual liberation and promiscuity. Sexual education should definitely include the fact that sex is best in a loving relationship and that most boys don't respect or want a girl who will have sex with anybody. A new study published while I was writing this book found that teaching abstinence when combined with a detailed and comprehensive sex education does work, delaying the onset of sexual activity well into the late teens or early twenties. The fact one should always remember about teenage sex is curiosity kills the cat and the more ignorant teenager are about sex the faster curiosity kills the cat.

Some things are commonly judged as immoral and bringing damnation to those who do it, such as homosexuality. Some men are born queer. Their brains have the same proportions as those of a female brain. In New York early in the AIDS epidemic the coroners learned how to tell the different between the IV drug users and the homosexuals just by examining the brain.

Lesbians are often the victims of sexual abuse in early adolescence by a stepfather or older male living with their mother. They come to see sex with a man as a betrayal of their mother and sexual contact with a male disgusts them. They are victims, whose normal sexual development was derailed by this abuse. Homosexuality is not a choice it is almost always the result of circumstances beyond the individual's control

Abortion - While a fetus is human life, a woman owns her own body and there is no easy answer to the issue. I am old enough that I knew a woman who died of a coat hanger abortion, before abortion was legalized. Many people think abortion was always illegal in the United States, before the Supreme Court legalized it, but abortion was only made illegal in the 1930's. Abortion plagues American society as a complex and thorny moral issue.

A fetus is alive and learning from the first few weeks of its existence. Ultrasound pictures of a fetus only a few weeks old show that the fetus swims inside the womb. Take a baby who was recently born and place it in the water, it will safely swim just like it did in the womb when it was only a few weeks old. When babies learn to crawl, they are also drawing on the lessons of swimming in the womb.

Yet it takes one hell of a lot of moral arrogance to tell a woman she must bear a child that she cannot feed, clothe, or otherwise take care of. Adoption after a woman has borne a baby for nine months and given birth to that child is an unnatural act. The pregnant woman bonds with the baby inside of her. The woman is conscious and aware that she is bearing a new life, separating herself from that life is difficult.

There is a question about abortion most people don't consider. Is human life sacred? The state has the right to take human life under the law. A criminal may be shot to prevent a crime or an act of violence. Soldiers may die when they are sent into battle to protect the interests of the state and war is really organized murder. Under the law an individual has the right to take human life to prevent death of or serious bodily harm to themselves or another. If human life was sacred everybody who ever took a human life in the course of their lives would be automatically damned and I do not accept that.

Most people think the law and morality are one and the same thing. They are not, morality does not come from the state. The law regulates public behavior, and maybe bent to serve an immoral purpose. In Nazi Germany the treatment of the Jews was the law. In the United States the forced sterilization of free citizens was at one point the law, and you cannot forget that the segregation and Jim Crow treatment of the blacks in the southern United States was the law for almost a century. The state derives its authority from the consent of the governed and democracy is based on the rights of the individual. Morality and moral choices are different in every individual. The state may regulate public behavior, but nobody stands in judgment before God for the actions of another. We are each judged based on our own behavior and the choices we make are our own, not somebody else's.

The standard I use is this: is the outcome of the act good or evil? So long as the person is not hurting others don't worry about their choices. They will be judged and you are not God. You can say this doesn't apply to abortion because the mother is hurting the fetus, but at what point do you say the mother no longer owns her own body. Even if you say it is human life from the moment of conception how do you choose between the rights of the mother (the life in being) and the rights of fetus, (the potential life within)? That was the original standard of Roe versus Wade the Supreme Court case, which legalized abortion in the first place. The court chose the right of the mother to make decisions effecting her own body as a right of privacy over the rights of the fetus to life because the fetus was not yet a life in being. The court has modified this doctrine to the point that the state now has the right to ban partial birth abortions even when the mother's life is threatened. (Partial birth abortion is where the child's brain is vacuumed or sucked out of the fetus' head after the fetus is partially extracted from the womb into the vaginal birth canal) There are rare cases when the physical act of giving birth will kill a woman. The Supreme Court has said a medical procedure could be outlawed regardless of the circumstances in which it was done, regardless of who it hurts. It was a decision that may one day cost a woman her life

All in all, human behavior is quite complex and morality and moral choices are equally complex. You cannot simply say God wants you to do this or that. You cannot dictate morality and human behavior. Everybody's choices are different and you must carefully build morality in individuals and society at large. It is sheer folly to preach that you know what God wants of everybody in a modern society. You can build morality in society through faith, but faith must be tempered with doubt or else it becomes a fallacy of the intellect more dangerous than any subversive philosophy.

Morality is quite complex and cannot be rigidly imposed on every individual in society. Faith is important but you cannot judge people and different societies based on a rigid doctrine of God. It is faith tempered by doubt and understanding of the complexity of life and moral behavior which must serve as the well—spring of our human existence if we are to lift ourselves above the purely animal nature of our origins and succeed as a species.

Rational Faith in a Technological World

"Yes, but we are here."
The reply of my Muslim friend Ali al Marsi to my remark
"Mohammed was right when he said, all is vanity".

The purpose of this book is to make people think. The major appeal of the Christian religion is emotional. It fulfills a need inside of us for identity and belonging. Creationism holds that humanity is special, unique and God's crowning achievement. Consider the infinite and eternal Universe carefully. Understand that the universe has no beginning and no end, in size or time. Then the Universe is an infinite engine of life and death. Even if there is life in only one in a million of all stellar systems, there is still an infinite number of planets that support life. Sometimes it is life, like the dinosaurs. Massive animals that roam the planet, eating plants and each other, animals who are no more intelligent than the average cow. Sometimes this life is like the porpoises, highly intelligent with language but lacking the hands to build anything and finally sometimes this life involves intelligent tool builders who have language like us. These worlds and life forms exist for eons before they die a cosmic death. Humanity is not the only intelligent

OF SCIENCE AND GOD

species on the planet, much less in the universe. If the Universe has no beginning and no end, in size or time, an infinite number of intelligent species have drawn the breath of life and died a cosmic death before our solar system even came into existence.

If you think God created the Universe some 15 billion years ago, ask yourself the question what did God do the first few trillion years he was hanging around? (and two or three trillion years is not even one millionth of one percent of eternity). If the Universe has no end in size, where is heaven? If you maintain heaven exists in another dimension, how do you assume a three dimensional body into it? Since there are other gospels written by the rest of the apostles what makes one right and the other wrong? If mankind was God greatest creation and always focus of God's plan for the earth, why did he give the dinosaurs dominion over the earth for 170 million years? These animals lived when there were no humans on earth. If God gave humanity free will how is Jesus going to save the world? Doesn't the world have to save itself? If you say Jesus lived over thirty years and never had sex, doesn't that make him abnormal? Ask yourself, what are your assumptions? Can you support them? Are your views based on rational thought or emotional arguments?

I submit to you that the entire human race evolved from a lower form of animal and we were not specifically created by God. Yet as creatures of God, we are the dominant life form on this one planet, in an infinite number of planets in a universe without end.

To me God cast his gaze upon humanity 2500 hundred years ago and preordained what will come out the next several thousands of years. He began the human quest for meaning by ensuring Buddha, Socrates and Confucius lived and taught their doctrines, giving birth to philosophy. God may have planned the struggle for the soul of humanity as far back as Moses and the burning bush or maybe in staying the hand of Abraham, ending human sacrifice. We ultimately are nothing before God. Yet, we are here and God has acknowledged our existence through Christ. God's acknowledgement of the worth of humanity is implicit in any healing miracle and through Christ's healing miracles God endorsed the doctrine and ministry of Christ Jesus.

In the superstitious world of two thousand years ago any man who could heal would be thought of as divine. Considering that it is truly amazing that all the Apostles did not contend Jesus was divine. Amazing that the Gnostic Apostles up held and defended his humanity. In reality no man or woman could heal without a divine mandate. Jesus may have ultimately been human, but he was truly a Christ, someone anointed by God. His healing miracles served to certify his doctrine, of forgiving others to be forgiven, judge not lest ye be judged, do unto others as you would have them do unto you and turn the other cheek. If you can find a better doctrine for human relations, I would like to hear it. Christ's doctrine upholds all of the higher instincts of humanity and does not lend itself to any of the baser motives and instincts of man.

Some of Jesus' doctrine is difficult to adhere to. "If a man strikes you on one cheek, offer him the other." This is a formula for being beaten to death by a madman but consider its application in the non-violent movements of the Mahatma Ghandi and Martin Luther King Jr. which freed India and won equality and civil rights for America's black citizens. Back in the seventies some people developed a moral doctrine of situational ethics, where the supposed solution to every situation was based in love. But there is such a thing as tough love and should one tolerate being taken advantage of? Returning love for abuse is ultimately a losing proposition. Turning the other cheek assumes the man striking you has a conscience and will tire of beating you.

OF SCIENCE AND GOD

There are other parts of Jesus' doctrine that have unexpected applications. Jesus said "Verily, verily I say unto you if you speak ill of your brother, you are in danger at the judgment." Clearly this means you don't go around calling people niggers, kites, faggots, spics, or other racial or ethnic slurs. Muslims take this to heart and are better at living by this rule than are most Christians. Most people don't realize it but this doctrine also means you should be tactful and lie sometime. If you know a homosexual and you are asked, "Is he a faggot?" The correct response is "No not that I know of." If you have had sex with a woman and someone asks you, "Did you screw her?" The correct response is "No" You don't say "Yeah, she is a slut" Being tactful requires you lie. The correct response when someone asks you a demeaning question about someone else is to be tactful and lie. Like the Bible says, there is a time and a purpose to everything under heaven. The Bible doesn't specifically say so but there is a time to tell the truth and a time to lie.

So, it is not by blind obedience to the doctrine of Christ by which individuals manifest their faith in God, but rather through the thoughtful, carefully considered application of Christ's doctrine by which mankind advances morality and the common good. One should not make the mistake of believing the proposition that, living by the moral doctrine of Christ is always its own reward. Some of the time it is harder and more demanding than an immediate selfish response, but it ultimately lifts up society as a whole. The pay-back is not immediate and seldom fits the needs of instant gratification, but in the long-haul Christ's doctrine of forgiveness, judge not lest ye be judged, do unto others as you would have them do unto to you and turn the other cheek is the best medicine for society and social improvement. You don't have to give endlessly. You don't always have to speak the God's honest truth no matter what. In point of fact, protecting others means sometimes you lie. Obviously you don't always lie, but the very least you can do is live by the saying. "If you don't have anything good to say about someone, don't say anything" It is not easy to always bite your tongue. But it is the Christian thing to do.

Ultimately the two principles of Christ's doctrine which give people the greatest personal return, are forgiveness to be forgiven and judge not lest ye be judged. Demeaning others and judging others is an endless task. The self-righteous always see wrongs to be corrected and cannot rest while this or that perceived injustice remains uncorrected. The truly righteous are calm in the face of injustice and adversity, while the righteous work to correct social ills. The self-righteous can work themselves into fits of passion over this or that perceived wrong and are never happy with their situation. They see the path to happiness in correcting others and in their version of social justice, never in themselves. People who are self-righteous are commonly blind to their own sins. The more you accept yourself and others as flawed, the less judgmental you are, the more forgiving and tolerant you are, the more likely you are to be content.

What do you gain by judging another person? How do you profit from morally condemning others? Remember the chapter on the abuse of faith: moral condemnation can be a vain and selfish thing. People who publicly denounce the immorality of others are either seeking to bolster their own ego by self-righteous pronouncement or are trying to agitate others. The great strength of the Catholic Church is that its priests learn to be forgiving and non-judgmental while providing moral solace.

Consider nymphomaniacs they may screw four or five different men a day. They however hate men and do not have orgasms. They are psychologically incapable of experiencing the ultimate physical pleasure of sex with a man. Nymphomania is a psychological disorder caused by frequent and repeated sexual abuse during the age of five to ten years old, the years of moral development. They were and are victims. Most people call nymphomaniacs whores, but in reality, they are sick, psychologically disturbed. What are you accomplishing by calling a nymphomaniac a whore? nothing! You are demeaning them and saying they are worthless. How have you improved yourself by doing this? Consider them sick, do not have sex with them and speak no ill of them and you are a moral man. If you want to save the world, try it one person at a time, one small act after another. It is the sort of humility and human decency careful consideration and application of Christ's doctrine can bring.

The principle I find most important in life and behavior is, do unto others as you would have them do unto you. When Jesus was asked which commandment was most important, he answered the only commandment is, "To, do unto others as you would have them do unto you." This was Jesus' commandment. All his other teachings and doctrine can be said to flow from this one principle or commandment. If you unintentionally wrong somebody, don't you want to be forgiven? We all make mistakes. Do you want to judged solely based on your mistakes? Forgive to be forgiven. We all do different things, different ways of having fun, have different ways of living life, make different choices or practice ways of believing or not believing in God. Do you want to be judged by someone else's standards? Judge not lest ye be judged.

Turn the other cheek is a little more complex. What purpose does violence serve? Isn't it better to be slow to anger or slow to fight? Isn't non-violence a better instrument for social change than terrorism? If you have done nothing wrong, do you want to be killed for minding your own business, so why kill someone to advance your cause? Is the use of violence a better method to change your adversaries than shaming them? In the long run, wouldn't you rather be shamed than badly beaten or killed. Obviously if you step into a boxing ring you don't turn the other cheek, but in life if you want people to help you, help others. It all flows from, do unto others as you would have others do unto you.

The solution to the world's problems is not the adoption of any overriding moral doctrine, but rather suffusing human society with a basic doctrine of fairness, self-reliance, tolerance, understanding and forgiveness. The constant criticism of different societies, social institutions and other people benefit no one. Accepting social disagreements without all the self-righteous rhetoric would lead to a better world. The bottom line on faith in the modern technological world is this: if you believe that there are things in the realm of human experience which science cannot explain, then you have a rational basis for faith. It is not so important in what terms you think of God. Morality flows from the conviction that there is a God and you are judged. This can be a source for moral behavior in the modern world.

Judgment

*"They are like chaff driven by the wind; Therefore,
the wicked will not survive the judgment"*

-from the first Psalm.

Live in the certain knowledge that you are judged. But what are you judged on? How much credit are we given for being fallible? Do you have a moral compass? Jesus said "If you murder your brother, you are in danger at the judgment." He did not say you are automatically damned if you commit murder. Most people don't realize it, but the Mafia has highly developed morality, a very strict morality. For example, screwing someone else's wife will get you killed. Everybody makes choices in life, some of us make hard choices. If you join the Mafia or associate with them, you are choosing an insiders' life where you skirt the law and society's rules to win the acceptance of a select few. You also accept that if you screw up you will be killed. Human life is not sacred in the Mafia ethics. If you join the Mafia, you will be expected to beat people, to use brutality and violence to advance the interests of your group. It is a very cold and calculated choice to make in life.

Let us compare two men: President George Bush and mafia member Sammy "The Bull". George Bush ran for President of the United

States because he wanted the easy life style and status of being president. He had no real program or platform. As soon as he assumed office, he started agitating to get rid of Saddam Hussein, because Saddam tried to kill his father. After 9/11 he invaded Iraq and started a war that has killed over 5,000 Americans and tens of thousands of Iraqis. If you provoke or use violence it is a factor in your judgment. While you cannot blame George Bush for the sectarian violence, he still has the blood of thousands on his hands and that is part of what he will be judged on when he dies. All because he could not forgive an act of malice against his father. Love your enemies?

Consider the case of Sammy "The Bull", a member of the John Gotti Mafia family. Sammy "The Bull" killed eight people in his career. The FBI bugged his office and in two years never gathered one shred of evidence from this eaves dropping. Sammy "The Bull" was however arrested because John Gotti had a fat mouth and the wiretaps of his clubhouse supplied volumes of evidence against many in the Gotti family. Faced with a long prison sentence Sammy "The Bull" contemplated murdering Gotti and taking control of his Mafia family. When he made up a list of people, he would have to kill to gain control the list totaled over a dozen people. Sammy "The Bull" decided it just wasn't worth it and decided to avoid prison time by turning states evidence. George Bush started a war in callous disregard of the human cost. Whereas Sammy "The Bull" made the decision to become an outcast and risk death rather than murder more people. Is Sammy "The Bull" going to heaven before George Bush? Only God knows because God reserves the right to judge all men for their acts and omissions in consideration of all things both seen and unseen.

Confession is good for the soul. The first step in correcting any form of immoral behavior is admitting you did wrong. In theory the Priest represents God and in confessing, you are accepting responsibility

Of Science and God

for your behavior before God. You admit you did wrong and hopefully work to change the behavior. The CBS news magazine show 60 Minutes did a piece on John Gotti Junior who got out of the Mafia. During the entire interview he exhibited a nervous tic. He kept blinking one eye, not both, just one. He seems clearly conflicted about his past behavior and very badly needs to visit a priest and confess. The priest will not judge him and just talking about it will make him feel better.

How does the morality of Wall Street stack up to the morality of the Mafia? Jesus said "Man cannot love both God and Mammon. No man can serve two masters he will grow to love one and hate the other." The Mafia leaves people who pay protection alone. The mafia preys on people who engage in vices, such as drugs, prostitution or gambling. If you don't mess with the mafia, the mafia won't mess with you. In the recent 2008 subprime loan scandal, a lot of bankers went out of their way to sell questionable loans to people who weren't qualified and in the whole process destroyed the economy. They used a position of trust to lure people into irresponsible and bad loans. Why did they do it? Money. They earned commission on every loan made and deliberately made bad loans, just to get the commission. This was not just a few loan officers, thousands upon thousands of loan officers were driven by the wind and wanted their cut of the action. The practice was so wide spread it ruined the whole economy. In the long run they induced more misery than all the Mafia goons in this country, put together.

Much of sex falls into a category of being chaff driven by the winds, sex is seen as sinful because so many of us have so little sexual self-control. A woman who screws around is a slut, but a man who screws around is just a man. In reality both men and women who screw around think they are getting over on the other sex. Sexual promiscuity is more accepted in men, but is it any more desirable? So, few men have the sexual ethics of Paul Newman or Howard Stern, two famous American males who were both faithful to their wives. Many men and women screw around looking for a thrill they don't get in the martial bed. Few people understand the key to sexual satisfaction is not dependent on the number of different partners with whom you have sex with, but rather quality of the sex you have. The actor Paul Newman said it best, "Why should I eat hamburger when I have prime cut sirloin at home." If foreplay and sex last 15 minutes 95 percent of all women will orgasm. It is the quality of sex, not quantity of sex that matters. A sexually satisfied woman wants to please her man and is willing to be adventurous with her man. Too many people are chasing thrills, when the true treasure in sex is with a partner with whom you have built a loving and trusting sexual relationship. Monogamy is always preferable to promiscuity in all matters sexual.

You cannot say sex is just sex and harmless. Swinging is just chasing the thrill, chaff in the wind. Swing marriages last five years. They break up when the woman realizes that the man is really on a power trip and thinks he has a golden dick. I know of a swinger who set a dog on his wife. I'll bet he thought it was neat. To the wife's credit she left him the next morning. That is the kind of thing swinging leads to also.

Consider a pimp and his stable of prostitutes. A pimp convinces his prostitutes that he loves only them—that they can prove their love for him by having sex with other men. They will often deliberately get their prostitutes hooked on drugs as a means of controlling them. Pimping is a form of dominance that is straight out of hell. If blacks in society today want to understand slavery, they need look no further than prostitution and their nearest pimp. A pimp controls a stable of prostitutes but in reality, he is the whore—the morally corrupt one without a moral compass who debases himself and others.

Child abusers seek power over an innocent because they have no sexual self-confidence. They don't have the confidence to have sex with an adult partner, so they pick on and abuse defenseless children. There are many forms of sexual abuse, and sexual abusers are more often than not men. But I know of instances of nuns preying on little girls. There are instances of adult women, teachers and others who prey on young boys. Again, like the men these women lack the confidence to form a healthy relationship with an adult. Sexual abusers are weak people driven by a twisted and perverse desire driven by sexual urge over which they have no control. Chaff driven by the wind.

If you want to consider the soul of a man, who was considered moral, accomplished and well-regarded in society. Consider the soul of the German philosopher Hegel. He wrote a philosophy of master versus slave as the human dynamic and when he was done, he made the statement. "The French revolution burned Europe; the Germany revolution will burn the world." Adolf Hitler came along and turned this philosophy and evolution into the German Master Race. Fifty million died in World War II. Hegel may not have seen it all, but he knew enough to make his remark. Is Hegel damned? If you think so, beware the moral purity of thought. Remember Stalin's purges and the Spanish inquisition-both were attempts to inspire adherence to an ideology by torture and other coercive means. More than fifty million died indirectly because of Hegel, yet he knew and bragged about the conflict he helped create. Do you understand why Jesus considered vanity a sin?

Good and Evil is not a simple dynamic. Consider that Jesus said "I am the truth and the life." How many people have died in the name of God since Jesus walked this earth and was crucified? Consider the question. How much blood is on the cross of Christ? Turn Christ into God and you say he could count it unto the very last drop. I am not saying Christ is evil. I am saying Christ was not God and that God is beyond all human concept of good and evil. God judges all things seen and unseen. God judges and God damns.

God in Christ's doctrine offers a path to salvation. It is not simple or easy but it is the best guide we have. The doctrine itself, not salvation by election, which only pays lip service to the kind of behavior and attitudes Christ preached. Consider well that O. J. Simpson is supposedly saved by giving his soul to Christ and remember his remark "If I did it, she deserved it". Do you think God rewards that kind of arrogance? If you do, you are free to give your soul to Christ as your Lord and Savior. If you are an evil person, I don't think it changes a thing.

Salvation lies in living your life by the doctrine of Christ, not in Salvation by election. Do unto others as you would have others do unto you is an excellent foundation for moral behavior. You can do a lot worse than living your life by that rule. This basic cornerstone of moral behavior is complemented by the maxims: Judge not lest you be judged, forgive others to be forgiven and turn the other cheek and not responding to insult or injury with vengeance or violence. In laying out his doctrine for human behavior Jesus established a few simple axioms that lead to a contended, self-confident, positive and socially beneficial life. It is a basic morality that is valid and acceptable in all the major religions, Hinduism, Christianity, Buddhism, Judaism, Islam. What Jesus preached goes to the very heart of what is good and righteous in life. If you choose to think Jesus was divine, then accept his doctrine as holy writ and the true expression of the way God wants you to live your life.

If in heaven you were put on trial for being a moral person, would they call church officials to testify to your church attendance, belief in a specific religious creed or call people you knew and associated with to testify about your behavior. Think about O.J. Simpson and his remark "If I did it, she deserved it." He believes in God and thinks he is saved.

Do you think he is better, more moral and stands a better chance at the judgment than someone, who has committed no evil acts but doesn't happen to believe in God at all?

Belief in God may or may not make you a moral person. If you are a moral and good person, you are a moral and good person. How much evil is perpetrated in the name of God? We all have free will. If you reject the premise of God because of the evil done in God's name I submit to you that you are a better person, than anyone who commits murder in God's name or uses their believe in God to excuse or justify their own evil acts. God gives everybody free will and that specifically includes the right to not believe. God judges people based on their behavior, not their adherence to any creed or specific set of beliefs. Religion is full of people who claim that this creed or that creed is the only right and true creed. If just one creed is right, aren't all the others wrong. How do you choose the right creed? Doesn't it make more sense to say you are judged by your behavior, not what you believe in or adherence to any specific creed.

Jesus said you can judge a tree by it fruit. If you want to be found to be in good standing at the judgment raise good children. After World War II the University of New Mexico did a thirty-year study of child rearing. They discovered if you wanted your children to be moral and contributing member of society, all you had to do was love them. So long as you were not a pathological child abuser, it didn't matter how you disciplined them. You could send them to bed without their supper, shame them, use corporal punishment or, more simply calmly discuss their behavior and errors with them. Almost all children have problems and troubles, but so long as you honestly and truly love them, they would generally turn out okay. As Christ said you can judge a tree by its fruit. How you raise your children is a major factor in your judgement by God.

The truth is what you do, not what you or other people think. Life is the moral arena. You are judged by your behavior not by what you believe. There is no creed no matter how righteous that removes the responsibility you have for your own actions and behavior. If you believe Jesus climbed up on the cross to offer you a free ticket to heaven by invoking his name, without living by his doctrine you are a fool. Christ was anointed by God to heal in order to certify his doctrine and the whole of humanity is acknowledged and anointed through Christ.

Christ and his doctrine are a guide on a path that leads to personal contentment and grace. If you want to ensure that your love survives your death and the grave, live your life by the doctrine of Christ. Raise good children. Be forgiving of others. Judge not lest ye be judged. Be kind and helpful to others. Do unto others as you would have others do unto you. Save the world one small act at a time. What better testimony could there be at your judgment than "They lived a good and decent life and did no Evil." Try and let God see you in that light if you want to survive God's judgment.

The Second Coming

"Is someone leading us somewhere?"
from the song "Blinded by Science" by the rock band Foreigner.

If one accepts the idea that medical miracles are possible, proof of God and that God granted Jesus the power to heal to certify his doctrine, what is one to make of the second coming? Jesus said "I am the resurrection and the life." and the doctrine of the second coming of Jesus is central to Christianity. What is one to make of this doctrine? According to prophesy God will announce the return of his supposed son with famine, plague, war and death. The crowning achievement of Christianity comes at a terrible cost. The Muslims, Jews, Buddhist and Hindus and other religions are to some extent along for the ride. The Muslims under some circumstances maybe willing to accept the second coming. After all the prophet Mohammed used to say "Beware the second coming of Christ". If one does not accept the idea that Jesus is going to come in the clouds and everybody is going to fall down and convert to Christianity, what is one to make of the crowning achievement of Christianity?

I do by dream believe the soul of Jesus will be resurrected and restored to life in the body of a dead man, by a new Christ. This series of dreams formed in part the origins of this book. I wrote down my thoughts about religion as a way of organizing my thinking. I have debated about whether or not to put this chapter and description of these dreams into this book. Ultimately, I decided to include it, because it reflects on my beliefs. I am smart and grew up an atheist, yet I believe the second coming of Christ is possible.

The return of Jesus by his bodily resurrection by a new Christ, would like the Crucifixion, change everything and nothing at the same time. While according to religious doctrine the Crucifixion was the central act of Jesus' life, in terms of the day to day lives of humanity it changed absolutely nothing. In today's day and age, the second coming would certainly be news. Like men walking on the moon, it would in very short order become known to almost every person on the face of the earth, but it would change little or nothing in the day-to-day lives of humanity. People would still work, eat, sleep, defecate, have sex and women would give birth as they always have.

While many people will pronounce the resurrected Jesus as God, I am not so sure Jesus will go along. Don't confuse Christ and Christianity. Nowhere in the bible does Jesus say "I am God worship me." Christians exalt Jesus and worship Jesus as God. Jesus didn't ask for that honor and may not accept it upon his return, Jesus was a Jew and if Jesus returns as a Jew and affirms his Judaism, he will most assuredly renounce his divinity, throwing all Christian theology out the window. The return of Jesus is the return of a Jew and if Jesus reaffirms his Judaism it will lead to explosive growth in Judaism. If Jesus returns and reaffirms his Judaism within a few years there will be hundreds of millions of Jews.

If a resurrected Jesus does not renounce his divinity, it would almost certainly produce a war. The other major religions would be extremely loath to except this resurrected individual as God and would physically resist any attempt to make or proclaim a human being, resurrected or not, God. When societies clash war follows and even if a resurrected Jesus renounces his divinity there is a serious danger of war leaving us with the apocalypse.

Islam is the major religion which was founded and began after Jesus lived and died. How Jesus addresses and deals with Islam will be a major facet of Jesus' return. There is a lot in Islam Jesus would certainly like: the total obedience to the will of God and not speaking ill of your brother. The Muslim world is divided between Shia and other sects who bitterly oppose and fight each other. If one portion of the Muslim world embraces the second coming while the other rejects it, it will almost certainly lead to war in the Middle East. The times would be fought with uncertainty with lots of possible landmines for Jesus to negotiate around.

Like I said I had a series of dreams and I believe Jesus the soul of Jesus will be restored to life in the body of a dead man by a new and second Christ. By second Christ I mean a human psychic who can heal. In my dreams, there were four potential Christs all touching corpses. After the separate dreams of these Christs touching the corpses, there was one dream of the chest of the corpse opening up to reveal a beating heart. Followed by a dream of surgery to remove a bullet from the brain of the resurrected corpse and the revival of the corpse. I want to disabuse people of the notion I was dreaming my own future. Three of the Christs were women and the fourth was a black man I didn't recognize. I repeat I am a white man, so under no circumstances can any of the Christs be me.

Were the dreams wishful thinking? The first Christ was an Indian woman. A Hindu priestess whose role is to hug people every day. She dispenses thousands of hugs a day at a temple in India, a fact of which I was unaware when I had the dream. I had no idea who this woman was when I saw her in my dreams in 2006. Several months later I was watching a feature about religion on the history channel, there was the woman and again I was dreaming the future, something that is impossible. The other two women are true psychics known to me. The fourth Christ was a black man, whom I didn't recognize. So again, none of the Christs could be me. So, why do I mention these dreams? It is not my job to perform the resurrection and return of Christ to earth that I dreamed about. It is my job to arrange it. Hence my inclusion of the dreams in this book.

What would I do pursue these dreams? A second avenue presented to me in my dreams. A dream of a young Jewish girl thanking Pat Robertson or Pat Robertson's son (they look a great deal alike) for praying to Christ that she be healed of cancer. I imagine such a dream would come true if I led millions of Christians on the Christian broadcasting network in the following prayer "Christ, please heal this child as a sign that my dreams are of your return." The healing of a terminally ill child in response to such a prayer could only be interpreted as a sign that my dreams were true and that they were of the second coming of Christ, greatly increasing the public pressure to actually conduct the ceremony.

In 2017 after I had written an earlier version of this book. I prayed to God about these dreams. Prayed if I should pursue these dreams. That night I had a dream that Gonzaga University would go to three straight NCAA final fours without winning the championship the first two times. I took this as a sign I should not pursue the dreams until that happens. At the time Gonzaga was in the final four, but lost in the sweet sixteen next year, then they lost in the elite eight. The NCAA basketball tournament with its final four was canceled in 2020 due to the pandemic, a modern plague. Last year Gonzaga reached the final four but was upset by Baylor in the championship game. So, the fulfillment of the dream is still at least two years away at best.

The dreams took place in China. One of the dreams was of the removal of a bullet from the brain of the dead man. That is a distinctly Chinese method of execution. Getting the Chinese to actually conduct the ceremony will take a lot of doing. Why would they do it? Because they would expect a failure and would jump at the chance to discredit Christianity. The Chinese use criminals as organ donors. I have no doubt that if I can persuade the Chinese to conduct the ceremony that as the ultimate doubters, they will use a heart donor as the subject. This does not deter me. I am aware that God weaves healthy new flesh into the intricate fabric of a human body in healing a tumor. So, I find the impossibility of resurrecting a heart donor as more of the same impossibility of any miracle. The task before me is daunting. I cannot perform the resurrection, but I understand all the dynamics of it and have faith. All in all, it is just not as simple as Gonzaga winning a berth in three straight final fours.

It may never happen. The Chinese government may refuse to conduct the ceremony. Gonzaga may never make three straight final fours. Pat Robertson and his son may die, there are dozens of things which could happen to prevent these dreams from coming true, but at some level I believe it will happen, because by the very nature of the dreams, I am dreaming about the future and my predictive dreams of the future have been very accurate in the past. Which is a major reason why I believe in God. I am over 65 years of age and in my life, I have had less than 12 predictive dreams. To date I have never had a false predictive dream. I have had some strange dreams but my predictive dreams have almost always come true, a few of them remain unfulfilled. Paramount among the predictive dreams which are unfulfilled are the dreams of the resurrections of Christ. I may die and never live to see them fulfilled, but how and why did I dream of the Hindu priestess, six months before I had any earthly idea of who she was? It is ultimately a question of faith and I have faith.

Many will believe I am deluded, but a psychiatrist will tell you faith is not irrational in and of itself. You can't argue faith. Either you believe or you don't. The task before me is to build such a convincing argument the Chinese will risk it. I have put everything into this. Time and only time will tell if I succeed or fail. Understand that, and understand it well. It is a question of faith. Time and only time will tell, only time will tell, if my dreams will come true. Meanwhile, I build the case and hope for the best.

Another reason I write about these dreams is to warn people not to put God in a box. Prophesy comes true in strange and unexpected ways and Christianity is full of pronouncements about what did or did not, will or will not, happen that just plain don't add up. Seven of the twelve disciples contended that the resurrection was of the spirit and not the flesh. If so, is the physical resurrection of Christ yet to occur? If there is any validity in this argument, don't put God in a box. Expect the unexpected and trust in the idea that God will in time make his will manifest.

Even if my dreams turn out to be nothing more than wishful thinking—no more than my own mind seeking to resolve the conflicts between science and my own Christianity. Will you deny God's right to make his will manifest in his own way, outside of the bounds of common expectation? Do you think God is bound by the pronouncements of the humans who follow him or is he free to defy the common expectations of atheists, and adherents of other religions and Christians as well?

The people who exalt Christ as the divine son of God exalt

OF SCIENCE AND GOD

themselves for accepting Jesus as God and think they are automatically forgiven, that God will lift them out of this world and spare them the suffering that all humanity will experience. Jesus said two will be in the fields one will be taken the other will not. Strongly suggesting there is no rhyme or reason to who lives and who dies. It is what I expect.

So why pursue the dreams at all. I am not very happy about the idea I maybe a player in the death of millions and millions of people. But a Christian evangelist by the name of Jack Van Impe consistently maintained that Christ will live one thousand years after his return. That after the dust settles Jesus will bring a thousand years of peace. That makes him the messiah, doesn't it, a title Jesus claimed when he began his ministry. Van Impe claimed that after Christ's death humanity will live in peace for all eternity. It may happen. If it does the meek will inherit the earth. They will inherit the earth from the dead resurrected Christ.

There are tags on your strings of DNA that dictate how many times the DNA can replicate. If someone is raised from the dead, God could in the same act modify their DNA to ensure they live a thousand years. If God wants to kill a substantial number of the human race to eventually bring about peace for all eternity, it is God's to do, isn't it? I am not going to call for war, but I believe. Hence why I am not happy about it, If Gonzaga wins a berth in three straight final fours without winning a championship in the first two times and Pat Robertson or his son is still alive, I will pursue the dreams. The apocalypse be damned.

I want to say something about these dreams and my pursuit of the second coming. I pursue them but I have not entirely bought into them yet. The dreams maybe false, Gonzaga will again be favored to win the NCAA basketball championship. If they win the dreams are false. If Gonzaga again fails to win the championship and fails to make the final four in the third year the dreams are false. If a Jewish child is not healed in reponse to my prayers and the prayers of Pat Robertson and the CBN audience the dreams are false. Pat Robertson has yet to assure me he will grant me the air time for the prayer. I send him a pre-publication copy of this book, before Gonzaga made the final four and lost to Baylor and noticed CBN has changed its programming emphasizing miracles among its viewers. This would seem to indicate Pat Robertson read the book and was impressed by the arguments, but he isn't ready to commit yet. I am glad CBN has changed its programming to emphasize miracles among its followers. It is proof that healing miracles are not the exclusive property of the Catholic Church. I am encouraged but I am not willing to commit to saying the dreams are correct yet. Meanwhile I continue to talk to religious leaders and others in pursuit of the dreams and hope for the best.

What you expect God to do or not do, is ultimately a question of faith different to every person according to their own beliefs. All my dreams might be wishful thinking. Who knows until it happens? I specifically want to caution people not to make too much out of my dreams. I had them shortly after my faith was transformed by the miracle of healing of Kathy Toler mentioned in the chapter on God. The dreams might have been my powerful mind coming up with a version of the second coming which was acceptable to me. A million things have to go right for the dreams to come true and have the Chinese perform the ceremony. Even if these million things go right, a new Christ who is either a woman or a black man must actually raise a man with a bullet in his brain, from the dead. I am a white man so it can't be me. All of this is ultimate out of my control.

Like I said, all my dreams maybe my mind coming up with a scientifically acceptable version of the second coming. Religion is full of people predicting the end of the world—predicting this and that. I may go down as a great dud. If that is the way it turns out, so be it. I have been wrong before. I will be wrong again. I am not perfect. I am not high and holy. I don't think for one second that I am the new Christ.

I don't know what is going to come out of this. I don't think the second coming of Christ is going to automatically convert everybody to Christianity. Islam is not going anywhere. I think that the second coming of Christ is going to start a major war. A war that will kill a minimum of two million people. No one should fall down and worship me as some sort of savior. I think we are in for a rough ride. I am not bringing good news.

Don't make any assumptions about my image of God. God took a third of the Jews in the Holocaust. Did God do it to drive the Jews back into Israel for the Apocalypse? If Jesus comes back as a Jew and God restores unto the Jews what he took from them tenfold, in 100 years from the Holocaust, there will be billions of Jews. A third of the world might be Jewish. The Jews are hated and that is not going to happen without a major war in the Middle East. I am not bringing good news. I am putting it all out there. I can't control it and I don't care what comes of it. I also don't know all the answers to the many questions it raises.

We are all the Children of God. God has the right to spank the human race so we can grow up and begin to act like mature adults, living in peace and harmony. There are so many religions, governments and creeds all claiming to be the ultimate truth of humanity, overturning and replacing them with a new social structure promises to be an exceedingly difficult process. To me the second coming is God's plan for the social evolution of the human race. I am willing to accept my dreams and the selling of this version of the second coming as a task and duty before God for now.

I am not a saint, I am a sinner: I have taken human life and was guilty of extremely serious sexual msiconduct as a twelve-year-old. I do not love God. I fear God and his judgment. God granted a miracle to alter my faith and throughout my life He has shaped my belief by dreams. Dreams that were impossible but have none-the-less come true. I don't know if I am destined for heaven or hell. The question of my soul is "does God use men for a terrible purpose and then throw their souls away?" I will not know the answer to that question until I die and no one can answer it for me.

God has shown me something impossible and asked me to arrange it. I will try, knowing like my life itself, though it is not all good, it is not all bad. A path of trial and tribulation that leads to millions of years of peace for all humanity. To me God wants humanity to grow up and toward that end he is going to spank us. Not a pleasant prospect but something we must face up to and endure to move forward. So I am in the position of a young child getting his father the belt, that his father will use to tan his bottom. I am not pleased! I don't expect a great reward or a pleasant experience, but I am doing as instructed, with a sense of resignation and regret.

Everything in this chapter is the vision of one man. A flawed man, smart and imperfect man. A man who has a history of being wrong about things. Nothing is certain. The question of this book is what are the dreams of one man and one miracle worth? Whatever will be, will be. Time and nothing but time, will tell. Paul McCartney wrote a very beautiful song "Let it be." "In time of trouble Mother Mary comes to me, speaking words of wisdom. Let it be. Let it be." The song is very good advice. "Let it be. Let it be." Whatever will be, will be. "Let it be"

The science and arguments in this book may make me and my

version of the second coming famous. So be it. If I do become famous along with my version of the second coming. I want the song "Let it be" to become the theme song of my version of the second coming. If this idea of second coming appeals to you, you should do the following. You should listen to the song from time to time. If someone tells you, "The man is a fool, it will never happen" say "Let it be". If someone says "Praise God, Jesus is coming back" say "Let it be". Don't go out of your way to sell this version of the second coming. By all means spread the word, but if you endorse my arguments, don't try and win converts to your point of view. You can't argue faith. Either you have faith or you don't. You can't debate faith. When you are faced with criticism or doubt, simply say "Let it be." For now, I seek to publicize this book, my version of the second coming and wait on Gonzaga and the final four. Let it be. Whatever will be will be. Let it be.

The Challenge to all faiths

"Judge not lest ye be Judged"
-a saying of Jesus Christ.

The human race and the planet earth will still be here in a million or even a billion years from now. The dawn of man was not when we first walked upright or developed language, that was the birth of humanity. We are living in the dawn of man. A process began 2500 years ago with the birth of philosophy. Religion and philosophy define us, they shape the way we look at ourselves. Yet the various religions always seem to be in conflict with each other.

Examine the questions of life and break them down to the basics. The dichotomy between nihilism and faith makes these two truths self-evident.

One, without God and judgment there is no meaning or justice in life.

Two, without God and your own soul you are nothing more than an empty sack of chemicals endlessly interacting.

These two truths may be facts of our existence, but they do nothing to answer the questions posed by life. The answers to these questions, indeed the questions themselves are different in every language, culture and faith. All faiths must realize this and teach respect for those who believe differently. Ultimately the people who lead their respective faiths must ask themselves the question. Do they think doctrine is more important than human decency? All faiths have different strengths and weaknesses. There is no one true faith. Ultimately God created all faiths and all faiths must learn to live in tolerance of those with different beliefs.

THE ULTIMATE FATE OF HUMANITY?

"There will be ten virgins for every man in Paradise"
a saying of the prophet Mohammed

The sun will last for billions of years before it turns into a brown dwarf and ceases to warm the earth. Unless the earth is gobbled up by a passing star, the earth will still be here in a billion years. They talk about our oil and coal lasting hundreds of years. Whereas humanity must face up to and plan for a future that will last millions or billions of years. The future of humanity is not all that rosy and bright.

The world population currently exceeds 7 billion people. In the last century, they did a study and concluded that we can feed a maximum of ten billion people. They have since come to the realization that the world's oceans will be fished out by 2050 when the world population will be at least eleven to twelve billion people. The seas provide ten to fifteen percent of the world's food supply. We are looking at the prospect of mass hunger in thirty years. Next time you see a child under the age of ten. Take a long hard look at them and realize that if they live to be forty unless something changes, they are certain to live in a time of wide spread mass starvation.

We can feed eight to nine billion after the seas give out. That leaves a minimum two or three billion or a quarter of the world's population without much food to eat. Combine that with massive crop failures from scorching summer temperatures of global warming and you have the apocalypse. Vast sections of the world will be like Somalia of the eighties and nineties of the last century with collapsed governments and rival war lords fighting over food supplies. The wide spread poverty of the third world is already leading to massive migration into the more developed Northern hemisphere. If and when people can't eat, it will only get worse. This is a short-term problem. When the oil and gas run out, modern farming will break down, making an already bad problem worse. Toxic nuclear power plants which meltdown and infect the land with deadly radioactivity feed no one. How will we feed humanity?

Mass communications, the internet and cellphones are changing our view of the world. People think the human condition is improving. They see progress everywhere and feel technology will provide the solution to all our problems, but you can't eat a cell phone, television or computer. The wide-spread food crisis may lead to currently unforeseeable solutions. In World War I Germany suffered wide-spread food shortages and experimented with chemically produced food. "Ersatz" food, it was called. You can produce vitamins chemically. To feed the children in the currently starving and badly undernourished portions of Africa they produce a white paste. Eight ounces of which is the equivalent to a daily pint of milk and a megavitamin. Nobody is getting fat on it, but it saves many children from starvation. Will there be whole chemical plants producing something akin to library paste to feed billions?

Another possibility is using technology to build massive food grow houses. Somewhat like a pot grow house except on a much more massive scale. You can make forty to sixty thousand dollars a year by buying a small house, installing infrared lights and turning the house into a pot farm. They considered the problem of growing food specifically vegetables in space to feed colonies in space without dirt, in the book *Colonies in Space*. Their solution involves vast areas of racks with cloth or Styrofoam linings. The seeds would be placed on top of the cloth or Styrofoam lining and the roots would hang down below the lining in the open air and nutrients would be sprayed directly onto the naked and exposed roots, fertilizing the plants directly without benefit of soil. Lights above the racks would provide artificial light for the plants to grow. A structure the size of a modern warehouse with several layers of racks could easily grow dozens of times the vegetables as a farm of the same size with the roots fed to cows or goats for milk. MIT is experimenting with this sort of technology and it shows great promise. The problems of feeding the world are sure to grow worse during the next 30 years as the world's population grows with global warming, but they are solvable.

The next ice age is about 35,000 years away. The most logical cause of ice ages is that the sun goes through cycles. They have done neutrino studies of the sun and the fusion in the sun is only about a third of what is needed to sustain today's temperatures. The sun is heated by thermonuclear fusion at the sun's core. It takes a long time for the heat at the core to reach the surface of the sun. I have read figures as high as two million years. I don't buy that. The sun rotates and there are currents in the interior of the sun. Still, it takes thousands of years for the heat of the core or lack of heat at the core to reach the surface. The sun's total energy levels vary over time. The reduced fusion levels in the sun's core is just the next Ice age on the way. Right now, the sun's surface is hotter than it was two decades ago. Nobody understands the heat cycles of the sun, nobody.

During the onset of an ice age, over the space of a couple of centuries, the sun's surface may drop in temperature 30 or 40 percent. The seas will evaporate in lake effect snows, forming massive glaciers on the plains and savannas of the world. You can't grow wheat or corn on snow. You can't grow rice in freezing cold temperatures. Because of its high salt content, you can't use the land on the continental-shelf after the seas recede. The onset of an ice age results in a massive starvation of animals and mankind. Humanity will be in absolute chaos. Humanity has been around for 5 million years. That means we have survived over eighty ice ages. It will be rough and humanity will be damaged but humanity will survive the next ice age.

There is a cosmic threat to human survival which hangs over the world's head like a massive weight on a thin wet rope. The white dwarf campion to the star Sirius. This star is our nearest companion in the Milky Way—only being some four and a half light years away. The white dwarf companion is likely to novate at some point in the future. If and when the white dwarf companion novates it will exterminate most of the life on earth. The problem is not the explosion of the star. It is the radiation from the explosion and subsequent dust cloud which will envelop the earth decades later.

When the white dwarf companion to the star Sirius novates, it will give off a deadly burst of radiation. People in Hiroshima died from the radiation of the atomic bomb without any visible injury, burns or wounds. The radiation deaths occurred a couple miles from the epicenter of the blast, but didn't exceed ten miles from the bomb blast. A couple of miles and we are talking over four light years. But understand something and understand it well, that radiation that killed for a couple miles was from about three pounds of uranium. The white dwarf companion to the star Sirius contains billions of trillions of tons of matter, that will cook off in one massive thermonuclear explosion and it will come without warning.

The massive scale of what will happen is hard to grasp. Three pounds of Uranium produced deadly radiation for a sphere of several miles. The white dwarf companion to the star Sirius is larger than the earth and one teaspoon of matter from a white dwarf contains 250,000 tons, not pounds, tons, of matter mostly hydrogen and helium. It is all going to go off in one massive thermonuclear explosion.

To imagine the radiation, consider that one hydrogen bomb ten times the size of the bikini atoll blast detonated 100 miles above the geographic center of the United States would by its radiation kill everyone flying in an aircraft above 20,000 feet from coast to coast. People on the ground in skyscrapers and buildings wouldn't die, but people outside on the ground would get seriously ill from radiation poison. If you were on the ground in Denver and inside a parking garage you would be totally unaffected by the radiation, but your son out walking the dog would become sick from radiation poisoning and would in all likelihood die of cancer within twenty years.

Four light years is a long way, but if and when the white dwarf companion to the star Sirius novates it will be trillions of times worse than that theoretical blast. Four light years is about 20 billion miles. Sounds safe, doesn't it? Our blast which is taken from a military study on nuclear warfare and briefly recounted in a book on the development of the neutron bomb, was at an altitude of 100 miles. Divide 20 billion by 100 you come up with 200 million miles. A trillion is 5,000 times 200 million. So, the radiation from the nova will be five thousand times worse than the massive killer hydrogen bomb. When the white companion of the star Sirius blows, everyone in the line of sight of the star will die in minutes unless they are shielded by several inches of concrete or steel. If you are in a skyscraper opposite the line of sight from the white dwarf companion, you will survive the initial burst of radiation, but radiation from white dwarf novation will continue at a reduced rate for several weeks as the star expands in the sky.

How bad will it be for the earth? I do not know the exact position of the star Sirius in the sky. But I do know it is several degrees north of the terrestrial equator within the topics. That means all the people at the south pole will live. If it occurs during the winter months in the northern hemisphere the Eskimos will live. Again, a few inches of concrete or steel will save your life, but as the earth rotates the radiation will bathe all life above ground and that life will be sickened and die in minutes or hours.

The white dwarf companion of the star Sirius will grow dramatically in size, perhaps as large as the planet Venus in the night sky. For about two or three weeks it will give off a massive amount of light and be plainly visible in the day sky. If it occurs in a season when the star is normally obscured by the sun's daylight. If the nova occurs in a season when the star appears above the earth in the night sky it will give off more light than the moon.

Once the light dies down the danger from direct radiation will largely be gone, but the star will still be lethal or dangerous to life due to the neutrons or protons traveling from the nova at less than the speed of light. The radiation will arrive at the speed of light and without warning. The neutrons created by the explosion will quickly break down into electrons and protons. The protons expelled by the novation explosion of the white dwarf will travel at less than the speed of light and their arrival will be spread out over a time of months or years. The protons will arrive much faster than the massive amounts of dust which will in a few decades bury the earth. But their arrival will be spread out over time. When protons or neutrons pass through living flesh, they break down the chemical bonds in the flesh, killing cells at a massive rate. This is how a neutron bomb, kills people and animals, protons have a similar effect. The proton stream will affect everyone in the line of sight of the star Sirius. It will be extremely foolish to be out and about when the star Sirius is overhead in the sky, for years after the nova.

An odd note to all of this, the protons and radiation will not harm the stored grains and food stocks. All the wheat in the siloes. All the rice piled up in bags will still be edible. Even the meat, if it was slaughtered before the radiation caused diseases in the livestock, will still be edible. The radiation will have killed most of the bacteria. The meat will mummify and will still be edible. Depending upon what percentage of humanity survives, the survivors will not be in immediate danger of starving to death. Unless, that is, a sizable portion of the human race survives and the food distribution network and infrastructure break down. New York City and major metropolitan areas where substantial numbers of people live in concrete and steel structures will be difficult to sustain, but the railways will still function, being shielded from the radiation by the steel of the locomotive. So long as you are not out and about while the star is in the sky you will be safe. It will be surreal. Something straight out of a science fiction movie.

The radiation danger will pass, but decades later the dust cloud from the nova will envelop the solar system and will end life as we know it on the planet earth. It is hard to explain how massive this dust cloud will be. The entire land mass of the earth came from one super nova or smaller nova in the southern sky millions of years ago, before the earth rotated.

All the continents were one large land mass in the southern hemisphere. Then the moon struck the earth in the north hemisphere, gouging out a large chunk of the earth and giving the earth its tilted axis and spin. The molten core of the earth quickly reformed into a somewhat spherical configuration and the crust grew back. But the earth is still somewhat shaped like a pear. Fatter in the southern hemisphere than it is in the northern hemisphere. The southern land mass broke apart under the strain of the moon's impact on the earth and the continents began to drift north, turning as they slowly drift.

The star Sirius is near the terrestrial equator and in the tropics. Unlike the previous nova, which deposited its dust and debris in one large scab on the face of the earth. When the star Sirius novates, it will deposit its dust and debris on a large belt around the earth, evenly spread around the globe as the earth rotates. This belt of dust will be well over a thousand feet thick. Because the atmosphere will carry the dust the entire earth will receive some dust. A couple of feet in the poles. Over a thousand feet near the equator.

As all this dust reaches the solar system it will obscure the sun. Interference with the sun-light reaching the earth will drop the temperature into the range of one hundred to two hundred degrees below zero as the ash bathes the earth. The sun will turn red or black in the sky and the stars will be blotted out in the night sky. All the steel mills, all the power plants and all of the earth's major cities will be buried in this ash. Earthquakes will be common as all the ash begins to press down on the earth's plates.

Ocean levels will rise dramatically. As the ash begins to fill the oceans. All fish and sea life in the tropics will be wiped out by the ash clogging their gills and making it impossible to swim and eat the microorganisms that are the basis of the ocean's food chain. Some sea life will survive in the arctic region, but due to reduced temperature the sea in those regions will freeze over and the whales will die.

The sea in the tropics will not freeze over—at least not develop thick ice sheets, because the water retains heat and should stay above freezing. The north pole is covered in ice because the high salinity of the water permits the water to drop below freezing allowing a thin crust of ice to form. In addition, any ice that forms in the oceans of the tropics will quickly be broken up by the weight of the dust falling on top of the ice. So, the topical oceans should not ice over.

99.99 percent of humanity will be buried under tons of ash or drown in a vast deluge of muddy sea water that submerges their cities and homes as the oceans' levels rise. Between, the accumulation of ash and rising sea water civilization as we know it, will wiped out. How will the human race endure and the human race must endure?

We must build massive ships like super tankers and ride out the rising sea levels. These ships the size of super tankers will carry the surviving animals, livestock, food stores and plant seed, sufficient to ride out the decade or more of rising seas and frigid temperatures until the dust cloud clears the solar system and the sun returns to its normal shining in the day sky, the temperatures rise and the ocean levels stop rising.

A little like Noah's ark, these ships must be supplied not just to ride out the dust cloud, but also to reestablish life on earth as there will be only a handful of plants and animals that survive the catastrophe. The entire ecosystem will be destroyed and will have to be reestablished. Only a handful of people, far less than one hundredth of one percent will survive the catastrophe. All norms of civilization as we know it will be swept away.

What will the earth physically look like? A lunar landscape of ash. Devoid of life. The life will be on the ships. Wide swaths of land where the continents used to be devoid of plant life, surrounded by shallow seas. With the return of sun shine the basis of the ecosystem will be restored. With the sun will come evaporation of the seas and rain on land. The first step in restoring an ecosystem on land, is you spread the grass seed to the winds. Some plankton and sea life will survive even without help. Life is resilient and the basis of an ecosystem will quickly reestablish itself in one year's time once the sun resumes shining in the day sky.

How many people will survive and what will life be like for them? A couple of million people at most will be all that survive. You can't just save lives. You will have to provide for and carry with you on the ships the means to survive once the dust cloud has subsided. People reproduce much more slowly than the animals and it would be best to restore the plant and wildlife first. So, the survivors could flourish in a land of abundance. You will need to carry guns and barbed wire to hunt and drive the wild animals away from the crops. The barbed wire will also keep the domestic livestock away from the crops. We will return to the simple lives of shepherds, farmers and hunters. With one major difference, we will need to repopulate the earth. So, the vast majority of the people will be women. They will tend the flocks, till the soil and hunt the game.

Marylyn Von Savant, the woman who claims the title of the smartest woman alive writes a column "Ask Marylyn." She was once asked by a reader if humanity had to choose between the two sexes which should they choose. She chose men. Because they have both genes and could reproduce both genes and clone a woman. She is an idiot with no common sense. Men would have to grow women in test tubes—a risky proposition at best. Whereas women have the wombs to give birth, the male sex gene is a subset of the female sex gene and women could use genetic engineering to produce male ovum and more easily clone men. One quart of sperm without the seminal fluid could literally repopulate the earth, once you have women and the wombs to carry the fertilized ovum in. The smartest woman in the world doesn't even know her own sex.

The prophet Mohammed had a male ego. Ten women to every man would be nice, but the need to repopulate the earth dictates a higher ratio of women to men. More like a hundred women to every man. I feel any man with a robust male ego would agree with me. It's the women who will tone down the ratio, not the men. Women however, have to realize it will be armed women who will enforce the ratio and when pressed for survival people will do almost anything.

There was an American Indian medicine man who started the Ghost dance, in the belief that the white man would sink into the earth and the buffalo would return. It may happen. The American buffalo are a more efficient grazing food source than cattle. It would make sense to take them over beef cattle and the Plain's Indian culture and life-style would be easy to recreate. We are not going to live in cities and drive cars. There is a line in the song Woodstock: "We are star dust, we are golden and we have to get back to the garden." Eventually humanity will return to the garden of Eden and live simply as we lived for millions of years. It will be paradise. Our culture will shape the planning for the world after the nova and the resulting catastrophe. But eventually humanity will return to its essence, without the trappings and pretense of civilization.

When will this happen? It will happen one day but nobody knows when. The knowledge and technology to predict it simply doesn't exist. Some people will read this book and say G is dropping it won't happen for millions of years. The case however is not that clear. It involves the speed and distance of the white dwarf's orbit from the star Sirius. As distance between the star and white dwarf expands as G drops, the orbit will slow down. The slower the orbit, the more pronounced the danger that the white dwarf will bleed off enough hydrogen and helium from the star to ignite in a thermonuclear reaction and as G drops the star will expand in size making all of this more likely. It is not a question of if it will happen. It is a question of when it will happen. Could be today could be tomorrow, maybe it won't happen for millions of years.

We can have some warning. Astronomers can study the white dwarf with the Xray observatory. A dramatic increase in the X-rays emitted from the white dwarf due to it drawing off hydrogen from the star Sirius will give us some warning. Know that nobody knows what level these emissions will reach just before the nova. The scientific knowledge simply does not exist to accurately predict when this white dwarf will novate.

I repeat when the white dwarf does novate, the radiation will arrive without warning. The dust clouds will take years to arrive and we will have time to prepare for that. We can take to ships and survive that, but will technology fade away into the night? We will still need metals for tools. You can't fit a blast furnace onto a ship. We will have to take basic knowledge of our technologies with us so we can recreate the industries to build the tools of survival as the population expands and grows. The most demanding technology is our computer chips and information technology. That requires special factories insulated from vibration. You simply can't build a computer chip on a rocking boat. It can't be done. These are some of the problems that will confront us.

Will religion fade away with technology and civilization. Maybe, maybe not. With no synagogues, mosques or churches, will people still read the Bible, Torah and the Koran? Will the doctrine of Christ survive? Will the practice of religion take a back seat to the needs of survival? Thanking God for your survival after a catastrophe is basic human nature. Surely these arks of survival will include ministers, priests, mullahs and rabbis. It becomes a question of what the survivors will teach their children about God and what will the children teach their children. I feel certain that humanity will still believe in God, but no one knows what shape or form that belief will take.

We will go back to live simply and in tune with the rhythms of nature. We will go back to the garden. It will be Mohammad's paradise. I have no doubt we will be happier and more contended in the fate we have drawn. What is humanity's essential relationship to God devoid the trapping of civilization? Time and only time will tell and answer that question for all humanity for all time.

Know there is a God who judges you,

so do unto others,

as you would have others do unto you.

True faith can be that simple.

www.ingramcontent.com/pod-product-compliance
Lightning Source LLC
Chambersburg PA
CBHW020759160426
43192CB00006B/375